Praise for *Water Wars*

"This is a wonderful book, a wake-up call of startling clarity and insight, with a flood of facts and anecdotes that place the abstract into riveting human perspective. I will never turn on the tap again without thinking about where the water comes from and where it goes."

—Ken Burns, producer and director of the
Civil War, *Baseball*, and *Jazz* documentaries

"Critical decisions about water . . . will have to be made in the future. *Water Wars* can help us make those decisions wisely."   —*BookPage*

"Clear, jargon-free . . . *Water Wars* chronicles not only ambitious construction projects but also the ambitious personalities behind them."

—*The Baltimore Sun*

"Famed as a water planet, Earth has too much in some places, too little in others, and everywhere the crisis of matching water with people. *Water Wars* is a brisk and personable introduction to a once and future struggle that is not going away."   —Stephen J. Pyne, author of
How the Canyon Became Grand

"*Water Wars* is a . . . wonderful book. I do not know Diane Raines Ward, but I suspect she is a solid journalist with a great thirst for knowledge both broad and deep. She spent ten years circling the globe to beautifully describe many of the world's most fascinating water development problems and projects and the people who masterminded—and mismanaged—them. . . . Splendid insight . . . thoughtful . . . informative."

—*Environmental News*

continued . . .

"No matter what your stance is on various water management issues, Ward's descriptions of the mechanicals of control are fascinating. . . . [Her book] offers provocative insights on old and new problems. . . . She's also not afraid to travel, to learn and talk to the kind of people whose work shoes are rubber irrigator boot, all of which adds a refreshing spin. *Water Wars* is . . . both a good primer on water management and an enlightened, engaging analysis of its woes."

—*Las Vegas Mercury*

"Diane Ward describes the competition for this scarce resource vividly, and with remarkable balance. A book for everyone to read."

—Robert O. Collins, professor of history emeritus,
University of California, Santa Barbara, and author of
Documents from the African Past and *The Nile*

"For years we worried over not having enough food to feed the hungry and enough oil to power civilization. Now, says Diane Ward, in this eye-opening investigative book, we'd better come to grips with an equally stark reality—that we are running out of fresh water. *Water Wars* presents a panoramic look at the next great crisis looming on the horizon, and what we need to do to address it. It should be read and widely discussed while there is still time to change course."

—Jeremy Rifkin, author of
The End of Work and *The Hydrogen Economy*

"An engaging story . . . Many of the twentieth century's greatest victories are double-edged when it comes to water. Ward talks to countless people caught up in the contradictions . . . at once compelling and sad [with] plenty to inspire and alarm." —*BusinessWeek*

"A warning about the worldwide struggle to manage water resources in an era of growing demand and climactic instability . . . [Ward] pursues a far-reaching itinerary in order to evoke the global nature of the crisis. . . . An informed discourse about the vital historical relationship between humans and water, and an overview of a possible global dilemma." —*Kirkus Reviews*

WATER WARS

*Drought, Flood, Folly,
and the Politics of Thirst*

DIANE RAINES WARD

RIVERHEAD BOOKS
NEW YORK

Riverhead Books
Published by The Berkley Publishing Group
A division of Penguin Group (USA) Inc.
375 Hudson Street
New York, New York 10014

First Riverhead hardcover edition: August 2002
First Riverhead trade paperback edition: June 2003
Riverhead trade paperback ISBN: 1-57322-995-4

The Library of Congress has catalogued the Riverhead hardcover edition as follows:

Ward, Diane Raines.
Water wars: drought, flood, folly, and the politics of thirst / by Diane Raines Ward.
p.   cm.
Includes bibliographical references and index.
ISBN 1-57322-229-1
1. Water-supply—Management.   2. Hydraulic engineering.
3. Water—Political aspects.   4. Water rights.   5. Water-supply—
International cooperation.   I. Title.
TD345 .W257   2002   2002021301
333.91—dc21
Printed in the United States of America

10   9   8   7   6   5   4   3   2

For Geoffrey Ward, Garrett Ward, and Kenneth Raines
Husband, Son, Father

Contents

Introduction

Sweet Water

You never miss the water till the well runs dry.

TRADITIONAL BLUES

When I was a little girl, my family used to get its drinking water with a cast-iron hand pump that sat on our front porch. It was too tall for me to work the handle hard enough to bring up water from our well, but I remember that whenever I drank a glass, it was cold and good. Because of that pump I made a simple assumption: people everywhere had wells under their houses that held all the water they would ever need.

An embarrassing number of years passed before I thought much about where people got their water. On a hot day in North Yemen, climbing down a rocky path from the hilltop town of Kawkaban several miles to the plain below, I passed a dozen village women, headed straight up, carrying large, heavy pots of water on their heads. My Yemeni companion told me that the women made that trek every day throughout the year. "Allah go with you," the women called generously. I was shocked at how hard they had to work for something that, for me, poured at the twist of a tap.

I've since learned that my assumptions about people and water were way off. Forty percent of the world's population carry their water from wells, rivers, ponds, or puddles outside of their homes. More significantly, many do not have enough—1.4 billion, almost twenty percent of those living on the planet, don't have access to an adequate supply of clean water. This isn't because there isn't enough water to meet all of our needs. Human beings only use a quarter of the world's fresh water. But water is seldom found just where we want it to be or in desirable quantities. Most fresh water sits in glaciers or deep aquifers, or runs off into oceans, far from the demands of our sprawling civilization.

"If you end the oil supply, the motor stops," I was told by Turkish Minister of State Kamran Inan, in Ankara in 1989. "But if you stop the water supply, life stops." How well we use our accessible water is becoming more and more important to how we live on this planet. Managing water is at the very heart of life in dry places—from Los Angeles to Lagos, from Damascus to Australia's Murray Darling River Basin—where the margin between survival and disaster is narrow, and even small changes have an impact far more drastic than those in wetter, more secure territory. I have seen the words GIVE US WATER splashed in bright red paint over a village wall in the Khyber Pass; GOD BRING US WATER scrawled across houses in southeastern Turkey; and PRAY FOR RAIN painted on a truck in Texas, where a four-year drought has made the ground so dry that sparks from car mufflers start fires and water tankers are part of everyday life. Across the world, twenty-three thousand square miles, a chunk of land roughly the size of the state of West Virginia, turns to desert every year.

At the same time, an overload of water endangers other people and places. Venice is sinking as the sea rises, and Holland's delta is threatened as it never has been before. More than ten thousand people drowned in violent flooding in the Indian state of Orissa in late 1999. In Mozambique, after severe rains in the spring of 2000, catastrophic floods killed thousands and displaced half a million. Most of the world's people live in coastal areas or on floodplains, where they are increasingly at risk from floods, among the most frequent and most destructive of all natural disasters.

As humanity grows thirstier in the earth's driest places, or is threatened by flooding rivers and encroaching seas in wetter ones, the steady rise in population puts stress on the land and the water. It took all of history up to 1830 to put a billion people on the planet but only one hundred years to add the second billion. The third arrived in just forty-four years and the most recent billion came in a scant twelve years. Women are having fewer babies in many places and there is real hope that the population will stabilize in this century, but there are now six billion of us and we add the equivalent of a New York City to the planet each month. Ninety million people a year. Think of a million lives lost in a famine or war—those numbers are replaced in four days.

As our numbers swell, we use more and more water, dramatically increasing withdrawals from rivers and aquifers. Since the middle of the last century, while the population doubled, water use has tripled. At the same time, we have dirtied that water with human, industrial, and agricultural wastes. We now face the need to feed unheard-of numbers of people on the earth while at the same time accommodating the toll exacted by growing so much food—increases in fertilizers, pesticides, salination, deforestation, erosion, and overgrazing. More than half the world's major rivers are either polluted or going dry. Half the planet's wetlands were lost in the twentieth century, and freshwater systems all over the world are losing their ability to support human, animal, and plant life.

Each year my husband and I visit the Ranthambhore Project Tiger Preserve in the state of Rajasthan in India. The dividing line between public and private land there is dramatic—green landscape begins sharply at the forest line while the land outside the park has been cut clean. The trees have not been felled by monstrous, greedy people but by women searching for wood to cook their family's supper. In rural India, where wood is still the common household fuel, 837,000 acres of forest are eaten away each year. That treeless landscape haunts me. It represents people without choices. More than a billion people live in India, and it's become increasingly difficult for them to stay alive without devouring their land and water. As hundreds of millions of people move into India's middle class, they—like their counterparts in the developed world—will consume still more food, more heat, more energy, and more water, if they can get it.

In the tree-lined streets of New Delhi, I went to the Centre for Policy Research to talk with a man who has written a great deal about water, B. G. Verghese. India's thirsty millions weigh heavily on his mind. "Today the population is almost three times the size it was at the time of Independence," he tells me. "It's taken us three thousand years to get to a billion, but it's not going to take very long to add another billion. The consequences of this growth are already beginning to be with us.

"When the population was small, requirements were limited," he says. "It could be dealt with. You managed or you moved. Those options are no longer available. The overwhelming growth of our cities calls for the provision of water for life and for quality of life."

My husband, who lived in New Delhi shortly after India achieved independence, in 1947, tells me that there used to be low jungle around the pleasant neighborhood where Verghese's office is now. Jackals called at night; there were gray partridges in the bushes, and an occasional wild boar wandered across a front lawn. It takes hours of driving to find anything like jungle now. The growth of the city, like that of India itself, has spiraled out of control. Once leisurely and elegant, New Delhi, while still beguiling in many ways, has become an overstuffed metropolis, straining at every corner. New Delhi friends talk constantly of pollution, traffic, and illness. There are two Delhis, however, New and Old, and the old city is worse, choked with people and vehicles and plagued by persistent outbreaks of typhoid, dengue, and malaria.

A friend who lives in Old Delhi, Shadiram Sharma, says that while this city once was heaven, it is now most surely hell. Each day, two hundred million liters of raw, untreated sewage pour into the Yamuna River as it moves through the city. This situation is repeated around the country. "Fifty percent of India's morbidity is because of water," Verghese tells me. "If we don't have more water, cholera, gastrointestinal diseases, diarrhea, dysentery, malaria, skin diseases, eye diseases, and the epidemics we already have will all become more severe. There will be social and political incidents."

One of modern India's gravest concerns is its mounting need for electricity. The Indian subcontinent, struggling with growth and development, is one of the largest users of power in the world, and power shortages, brownouts, and voltage fluctuations are an everyday matter. In parts of Delhi, Old and New, the air is thick with smoke and noise from gas generators used to light houses when the city supply fails. But Verghese believes that however serious India's energy crisis is, the most frightening issue the country faces is its water shortage. "There are alternatives to energy," he says. "But there are no alternatives for water apart from recycling and desalination. And desalination is not an easy option in continental-size countries. Well, I've got water, so maybe my supply is all right, but what about the guy who has no water, what happens to him? Who's to be responsible? Will he say, 'Well, it's OK, my good friend Verghese has water. I shall sacrifice myself for him?' I don't see him saying that. Why should he?"

Verghese sees a further trap lying in his countrymen's faith in the waters that have sustained them so long. "The name Ganges evokes sustained civilization," he warns. "And if the celestial hosts could live off the Ganges water, why not we mortals?" Verghese insists that such faith fails to take into account the conditions necessary for survival, much less what is needed for a good quality of life or what is required to maintain the natural systems of the world. "When you are terribly poor, the environment comes down the line," he says. "Survival comes first." In 1985, 750 Indian villages were without any water source whatsoever. Eleven years later, 65,000 villages found themselves in the same predicament. Sitting in the midst of eleven million people, B. G. Verghese knows that without storage, without control, ever greater numbers of people are

going to be in crisis. "I don't think we can easily dispose of these matters," he sighs. "Rains are seasonal. They are not evenly distributed over
time or landscape. We have to plan for it. It can no longer just happen."

No one would contest Verghese's concerns about the heartbreakingly
poor people living just across the Yamuna River. He lives in a pleasant
neighborhood of a city already facing terrible water stress. I live in the
pleasant neighborhood of a city farther from those stresses, but I know
that B. G. Verghese is ultimately speaking for all of us: "The growing
pressure of this population is the core of the problem," he says, "and
anyone who doesn't look at this is being very, very shortsighted. The
next century is going to see a water crisis, and that crisis will be much
more difficult to cope with and much more complex than the energy crisis has ever been."

American environmentalist Joe Podgor agrees. "There is a fundamental ecological principle, a law of nature, called 'carrying capacity,'" he
explains. "A biological system can only provide so much food for the
protein-eaters, so much biomass for the grazers in the form of plant life,
and only so much water to go around. It's the law of diminishing
returns. You cannot invite an unlimited number of people to a dinner
table at which there are twelve place settings and one pot of soup. There
is a limit to growth. That's the carrying capacity."

We live within a system of finite resources. There isn't much we can
do to alter the actual quantity of water on the earth. In a finite liquid
cycle, the sun's energy sucks up water from the earth and sends it back
again as rain, sleet, or snow. But although this supply of water is largely
fixed in amount, a great deal can be done to alter its location and quality.
Our access to fresh, clean water has been radically transformed by
interventions—dams, storages, diversions, overuse, and pollution. The
effects of water use and misuse are, with increasing frequency, felt far
from their source. Those who believe the water problems of other areas
won't affect them ought to consider that water-short California produces about half of the United States' fruits and vegetables and much of
its dairy products. We should understand that draining the Everglades

has meant less rainfall for Miami, and that industrial effluent poured into the Rhine in Germany must be cleaned up in the Netherlands.

At the beginning of this new century, a third of all countries suffer water stress. Overpumping of aquifers has dropped water to critical levels from Athens to Osaka. In Bangkok, Jakarta, Manila, and Mexico City, groundwater has been so depleted that the ground underneath the cities is collapsing. In the American West, towns purchase land solely to gain rights to the water under it, and Los Angeles periodically hires "Drought Busters" to apprehend "perps" washing sidewalks or watering lawns. Taiz, a lovely city deep in the mountains of Yemen, receives piped water once every three weeks. Not long ago, eight people there died in a feud over a water well.

As the supply of water fails, its costs rise. In Onitsha, Nigeria, poorer householders spend almost twenty percent of their income on water. In Sydney, Australia, water theft, which can be reported on a twenty-four-hour hotline, carries a fine of $20,000. In Bombay, local mafias chain up the water taps and charge residents by the bucket.

B. G. Verghese's predictions about water stress leading to violent outbursts have already become a reality on his own subcontinent. In 1991, in Karnataka, eighteen people were killed and another thirty thousand displaced in riots protesting the government's releases of Cauvery River water. That same year, five thousand Bangladeshis rioted in Dhaka, smashing cars and stoning police to protest water shortages that they blamed on India's theft of water from the Ganges River. Recently in Tamil Nadu, armed robbers commandeered a train, holding passengers and security guards at bay, in order to steal the water out of the toilet tanks using buckets and cans.

Unrest is not limited to the Indian subcontinent. In some troubled spots, the means of controlling water—the works themselves—have been targeted by rioters or competing armies. In August 1998, in the Congo, rebel soldiers seized the Inga Dam, cutting off electricity and water to Kinshasa. Later that same year, rebels in Lesotho took over the Katse Dam Project, which was being built to supply water to South

Africa. In response, South African troops invaded Lesotho and recovered the dam, killing seventeen. In 1998, and again in 2000, Fijian tribal landowners carrying spears, axes, and clubs, fought off armed forces deep in the rugged center of the island Viti Levu to commandeer the country's only large dam and hydroelectric power plant, Monasavu.

IT'S NOT ONLY in faraway, developing countries where water shortages have sparked trouble. During the long drought in Spain in the 1990s, the dry town of Denia began planning a pipeline to a nearby river. Neighboring villages refused to allow Denia to cross their boundaries, and pipes for a water treatment plant were immediately destroyed by saboteurs. In Wales, several years ago, rural residents blew up two water-supply dams to protest the diversion of their water to the English city of Liverpool. It astonished me to learn that near my own—wet, I thought—hometown in western New York state, frustrated farmers with dry wells broke into public pipelines to get water for their crops.

In Callicoon, New York, a few years ago, residents organized a broad political assault on a neighbor who wished to sell the water that bubbles up out of the shared aquifer underneath their land to the Great Bear Bottled Water Company. Neighbors accused the man of drying up the watershed, driving out the black bears and raccoons, and wiping out their trout streams. In response, Paul Levy, a nearby landowner, posed a serious question: "Can anyone truly be given permission to sell ground or surface water simply because it flows through that person's property?"[1] This simple statement, made by one small landowner, is as relevant to nations as it is to neighbors. Who owns water?

There is not one internationally accepted solution to the problem of sharing water. There are, therefore, problems to be solved: how do we ensure that water is distributed fairly across watersheds; or decide whether water is a commodity or life's blood; or know whether or not wars over water are inevitable. As people and nations try to sort out their competing claims, the results can be deadly. A friend has given me a photograph, taken in 1903, of the corpses of Daniel, Alpheus, and Burch Berry, lying in front of a barn on a dry Kansas farm, three farmers shot dead over water rights by their neighbors, the cattle-ranching Deweys. I've thought of the dead Berrys often while writing this book. That century-old shoot-out symbolizes for me every kind of water battle, from farmers fighting over a lone well to nations contesting rivers across continents.

The question of ownership and rights has provoked countless disputes, from Slovakia and Hungary battling over the Danube to Namibia and Botswana challenging one another's rights to the Okavango. In the Canadian Magellan Islands, two communities furiously contest one fragile freshwater aquifer. In the United States, Georgia has threatened to call out the National Guard during a feud with Florida and Alabama over the Chattahoochee. Emboldened by its victory over Colorado in a $12 million lawsuit over the Arkansas River, Kansas then sued Nebraska to regain 10 billion gallons a year out of the Republican River.

Few sources of water are so insignificant that they cannot be a source of conflict. The federal government was recently called into a dispute between Yellowstone National Park and landowners who were tapping geothermal waters near it. The Park Service said these withdrawals interfered with Old Faithful Geyser. Even the clouds are up for grabs. Making rain in one spot can rob farmers downwind, opening up a whole new world of lawsuits. When eastern Montana farmers complained that North Dakota had been stealing their rain, the Montana Board of Natural Resources refused to give North Dakota a cloud-seeding permit. That judgment was later overturned, but the settlement stipulated that North Dakota had to pay for a study to prove that it wasn't poaching Montana's clouds. In another case, Idaho provoked Wyoming officials when it decided to seed clouds over the Grand Tetons in order to improve snow pack on the mountain ranges' west side. Wyoming wouldn't allow it because, the state insisted, seeding would create excessive runoff on the east side of the mountain slopes and overload local dams.

As we face the alarming fact that the physical supply of water has limits, we dig our wells deeper, remove salt from ocean water at huge expense, and use and reuse water over and over. We compete for control of shrinking rivers and move large amounts of water longer and longer distances around the world in stranger and stranger conveyances. Scenarios that may have seemed ridiculous only a few years ago—such as towing icebergs from the Arctic Circle south, to supply water-short cities—now seem plausible. Las Vegasites now shop for water across the Rockies in Wyoming, and representatives of sheikhs in Abu Dhabi have offered to build dams many miles and countries away in the mountains of Pakistan in order to shuttle that mountain water south. In Chile, clouds themselves are being harvested. At El Tofo, University of Chile researchers catch coastal fogs in great walls of polypropylene mesh nets, which trap moisture and collect enough clean fresh water to supply entire mountain villages. It takes ten million fog droplets to form a single drop of water, and yet a 40-by-13-foot fog trap can produce 45 gallons of water a day. Fifty such traps cull six gallons daily for each of the town's four hundred citizens.

The most ambitious water schemes seem to blur reality with science fiction. A Norwegian company has developed an enormous, collapsible, plastic-coated fabric container called a "Medusa Bag," which can carry 30,000 to 80,000 cubic meters of fresh water from the fjords to needy places like Gibraltar or Israel. Canada's Global Water Corporation signed an agreement to ship 5 billion gallons a year from Sitka's Blue Lake all the way to China using such Medusa Bags—until, that is, the Canadian government slapped a ban on bulk exports of water.

Our credulity is strained by increasingly fantastic proposals. In central Africa, engineers have proposed creating a vast lake to bring rainfall to the Sahel. One researcher at the Massachusetts Institute of Technology wants to store winter snow and water in giant mountains of ice by blowing water through snow-making machines and insulating the resulting artificial icebergs under vast Mylar sheets until the water is needed. But even real-life projects, such as China's three-canal scheme to pump water from the Yangtze River almost a thousand miles north to Shanghai, are hard to comprehend. Each breathtakingly ambitious idea follows hard on the heels of the one before.

WE HAVE ALWAYS thought big about water. With giant dams and canals, men move rivers, stop oceans, create massive lakes, make deserts green, and restructure entire regions. Some of our feats of water engineering are great successes—such as the polders and dikes of the Netherlands or thousands of years of irrigation along the Nile. They save lives, reclaim land, and enrich whole populations. Some have been disasters, muddying waters that once ran clear, their benefits never outweighing their cost in destruction.

Few things arouse as much awe as great bodies of water—rivers, lakes, oceans, and seas. Consequently our imagination is captured by the great engineering projects that seem to challenge nature itself. Massive, visually prepossessing, these monumental waterworks make us seem bigger than we really are, powerful beyond our puny individual abilities, more permanent on the face of the earth. They make us believe that we can control a force of nature.

History, however, is filled with stories of catastrophic mistakes that suggest such wholehearted trust in engineering solutions is misplaced. Political boondoggles, bursting dams, and errors in design, construction, and placement have claimed thousands of lives and brought about the suicides of engineers. The Vaiont Dam in the Italian Alps is one of the world's highest, a thin slice of arched concrete. When a violent landslide hit the back of the dam in 1963, the structure held, but millions of tons of water shot over the top, killing three thousand people in six minutes. The eight engineers who designed the dam were put on trial for manslaughter. Perhaps the most ghastly waterworks disaster in recent history occurred in 1975, when 230,000 people died after China's Banquiao and Shimantan dams gave way under heavy rains.

No nation on earth has had as disastrous a confrontation with water as the former Soviet Union. Because most of their water resources are in remote Siberia, Russians have been involved in a monumentally aggressive series of water schemes for forty years, building the world's biggest dams and longest canals but often failing to make them work as intended. In 1980, Russian engineers built a 1,800-foot canal to limit water flow from the Caspian Sea into the inland sea of Kara-Bogaz-Gol. It worked too well. In just three years, all 7,000 square miles of the Kara-Bogaz-Gol had dried up. Engineers then began work on a restorative aqueduct.

The most appalling loss anywhere on earth is the near obliteration of the Aral Sea over three decades. Once the world's fourth largest inland sea, it has lost two-thirds of its volume—15,000 square miles—mostly because of withdrawals of water from the Amu Darya and Syr Darya rivers for cotton irrigation in central Asia. If this process continues unchecked, the Aral Sea will soon be just a brittle memory. Even now, ships lie stranded in the sand and some former port towns are as far as 90 miles from the receding shoreline. The once thriving fishery is gone. Worst of all is the toxic nightmare that has devastated the lives of the million and a half people who live nearby. What water remains in the rivers that flow into the sea is filled with such a heavy load of pesticides, fertilizers, and salt that nearly all of the water—under or above ground—is contaminated. Lethal winds blowing over the exposed seabed carry salt and chemicals across the land, poisoning both the ground and the people. Little grows in the region, and the people suffer a plague of illnesses: kidney and thyroid disease, cancers, viral hepatitis, tuberculosis, and probably the highest rate of anemia in the world. Life expectancy is shorter by twenty years than it is in other parts of the former Soviet Union.

The stories of the best and worst of the big projects are often monuments to human folly, but they are also testament to ingenuity and persistence. Whether it has worked to our benefit or our detriment, storing and moving water or pouring it on plants seems to be as natural to man as staying in one place—we've been doing it for over seven thousand years. The pitfalls of hydraulic projects and development schemes are numerous, but we've been building dams, reservoirs, tunnels, and irrigation canals long enough to understand a great deal about how they work. While it became clear in the twentieth century that structural solutions may not be the magnificent remedy we once hoped for, engineering technology can be a weighty tool, if environmental and human dimensions are taken into consideration. To balance the needs of a sprawling civilization with a vulnerable water supply, we ought to carefully examine every potential solution.

It's important to take steps to ensure that our water is here for our children and our children's children, who, no matter how the birth rate has dropped, will live in a far more crowded world. First, we must grasp the fundamental importance of our water. "Science can affect the way that we approach a situation," political water consultant Joyce Starr said to me not long ago in Washington. "But I think science comes later, after people have understood what their water means to them . . . what a precious thing it is and that it's a dying resource. Yes, there is enough water. But the agencies that are trying to help are all overtaxed. As a world community we are actually standing by while thousands of children die every day from water scarcity or waterborne diseases."

Starr is right. Six thousand children die daily from water-related maladies. "Imagine the deaths occurring as we speak because we're not able to provide sufficiently for the people on this planet," says Starr. "The tragedy is in the eyes of the mothers and fathers, and in families broken by deaths and anger and hurt—this pain isn't carrying far enough into the halls of power.

"There is a tremendous gap between understanding and action," says Starr. "It's not only the men and women who lead nations but the interests behind them. Even if officials understand that people in their country will die, they may be powerless to change the realities of the powerful forces in their country. It takes a lot of energy and it's brutal. I think you have to throw yourselves up against the system. We have to be

Don Quixote. We have to, because the alternative is . . ." She shrugs. The alternative is unthinkable.

WHEN MY HUSBAND and I visited India in the early 1980s, I watched our friend Fateh Singh Rathore, then the Field Director of Sariska Project Tiger Preserve, confidently direct work crews changing the course of a river. On his orders, teams of men set about the arduous task of forcing the river into a new bed just under the Aravalli Hills. At first his efforts baffled me. Fateh Singh, a member of a warrior clan from a village where there is no perennial river, is a man of the desert. A man who had traveled on camelback through sand dunes to his wedding was altering a river's flow.

Since I first watched Fateh Singh at work, I've come to understand that people who live in places with little water are often especially canny about its ways. They need to be. For several years, we regularly returned to Rajasthan and observed the aftereffects of his work. The redirected river and newly dug waterholes, crafted to feed yet other waterholes via underground pipes, carried water more evenly throughout the preserve. Trees and vegetation flourished in a landscape once entirely arid scrub. Breeding patterns of wildlife changed, and populations of spotted deer, sambar, nilghai (blue bull), and the leopards that fed upon them increased. Moving water wisely changed the life of the Sariska Preserve.

While visiting Southeast Anatolia a decade ago, I watched earthmovers tear up the ground for Turkey's gargantuan, twenty-two-dam GAP project on the Tigris and Euphrates rivers, and became both fascinated and alarmed by the immense consequences of water control. It was evident that while those dams will bring Turkey's arid plains to life, they will also certainly deprive Syria and Iraq of fundamental water supplies. Since watching earthmovers plow through the Anatolian hills, I've talked to scores of engineers, politicians, farmers, builders, hydrologists, and conservationists to discover what they have to say about managing our water. I discovered that the questions and answers to our water problems are found around the world, some of them in unexpected places.

"The challenge is to get water where it needs to be and to stop wasting it," Joyce Starr said quietly. "As we speak, another lake is disappearing. Until you get a strong hold, philosophically, intellectually, practically, on the economics of survival, you can put Band-Aids on it but you can't resolve it."

As communities run out of water, new dams and canals will be built. When people riot in the streets over water, no cost will be too high, and concerns over consequences will go unheeded. "When you hear a number of people saying that something must be done, something must be done at once," said the imperial dam-builder Sir William Willcocks at the turn of the twentieth century, "you may be quite sure that something foolish will be done."

We can no longer afford foolishness. We are using our supplies of clean fresh water at a rate outpacing population growth. How well we manage the water we have is becoming a matter of life and death more quickly than we are prepared for. As pressure increases, the decisions we make need to be good ones. It's important to understand what works, and what does not and why.

Chapter 1

HOLD BACK THE SEA

Climate Shock and Rising Waters

Your foe Oceanus does not rest or sleep either by day or by night, but comes suddenly, like a roaring lion, seeking to devour the whole land. To have kept your country, then, is a great victory.

ANDRIES VIERLINGH, *TRACTAET VAN DYCKAGIE*, C. 1575

This is a strange country, an inverse country," waterman Pieter Huisman tells me. "In fact, there is no country!" Watery, boggy Holland, which looks deceptively like land and is home to some of the most pragmatic people on earth, is awash in contradictions. With four hundred inhabitants per square kilometer, the Netherlands is the most intensely populated place in Europe. Yet, if the Dutch were to stop pumping and unman their dikes, half of Holland would disappear. And that half is where three-quarters of the population of the Netherlands live.

Huisman, the head of the Flood Control Division of the Rijkswaterstaat is a pleasant-looking Dutchman, sandy-haired and earnest. In his suit and tie, he resembles a Shell Oil executive more than the sort of big-booted fellow you might expect to be identified as a "waterman." Unhampered by his business attire, he and I tromp around a sprawling area of ditches, canals, and polders not far from Den Haag, where he speaks of twelve different levels of water within five hundred feet. Disbelieving, I make him show me all twelve. The water moves steadily; it is pumped up, stored, discharged by gravity, pumped down, flushed, drained, and pumped up again. The complexity of maintaining the layers is staggering. "Every water level has its meaning," says Huisman. "Because the dikes and embankments are built in peat and clay, you have to control water level in order to maintain banks. If you have a drop in level, they crumble. Water will overtop inner embankments. You will soon have wet feet."

Huisman lives in Zoetermeer, a new town built on reclaimed land. When foreign guests arrive, he explains to them that the ground-floor living room in which they are sitting is below sea level. "This means that if the water comes they will also drown on the first floor." On the second floor, above sea level, he announces cheerfully, he keeps a canoe.

Huisman and I drive across the bottom of a lake reclaimed in 1642. We are thirteen feet below sea level, navigating roads on which houses cling dramatically to earthen banks lapped by waves. The Dutch are so confident in the men who control the water that they have built their houses on piles and bundled them close to canals—when the wind blows, water brushes their doors. In the very bosom of water, they have adorned their little dry houses with shutters, trees, rock gardens, and mailboxes, all signs of permanency—or maybe just hubris. Huisman tells me about an abandoned Roman castle. It lies under water four miles from shore, on land that the sea reclaimed from the Dutch.

Huisman, whose father was also a waterman, lived with his family near Dordrecht in the Dutch Delta, an area renowned for its capable engineers. He was twelve years old in 1953, when southeastern Holland suffered a serious flood. "I can remember every thing we did that day," he told me. "At half past eight in the morning, we received information that the dike was broken, and my father, who worked at the dike, said to us, 'Remove all our things from the ground floor and take them to the first floor.' We had a new stove. He said, 'Put it as high as possible.' We put it on a table. This was important—a stove was expensive in the fifties. People in the street thought we were crazy. They said, 'The dike is breached far away. How will the water come here?' Twelve hours after the dike breach, the water came and we were evacuated. We had water in our house but we saved the stove.

"My family had lived for centuries in that area and was aware of the danger," he continues. "But people who have not seen it don't know what can happen." Huisman, profoundly affected by the floods, worries about a generation that has forgotten the peril. "My cousins in the next generation say to me, 'Floods don't occur anymore.' Because of our dikes and protection works, people don't realize that they are living in a vulnerable situation."

When the Rijkswaterstaat recently began reinforcing dikes along the coast, the Netherlands increased its protection standards to "Delta Safe," which means a dike is able to withstand a storm of a magnitude that might occur only once in ten thousand years. Protection has become so good that no one under the age of forty has seen the sea wash over the land. Delta Safe has protected Holland's most densely populated and industrialized areas, which have subsequently become even more heavily populated and industrialized. "It is a paradox," Huisman says. "By strengthening the dikes you give greater protection, and so people settled in the protected areas. Increase safety and you increase the damage when a disaster comes. Now a disaster would be very great!"

In January of 1995, Holland's new generation got a taste of the old danger. Unusually warm weather melted large quantities of snow from the Alps and sent water rushing into the Netherlands, this time through the back door. The rivers that drain the Alps, the Rhine and the Waal, swelled to twice their normal size and rose high against the Dutch river dikes. When rushing waters push against a dike for long periods, the clay and sand walls become waterlogged. If the river recedes more quickly than the water held in the walls, pressure on soggy, weakened clay can topple the dikes. Holland was as ready as it could be. Military planes with infrared cameras flew low to identify trouble spots. Dutch soldiers worked alongside civilian volunteers to reinforce embankments with sand and plastic sheets. But two hundred thousand people were evacuated, along with a million and a half chickens, fifty thousand sheep, four hundred thousand pigs, and half a million cows and their milking machines. The Dutch barge fleet, the largest in Europe, was brought to a dead stop. Businesses lost about $84 million a day. Thirty people drowned, the hardest possible reminder of Holland's perpetual frailty. "We are still a vulnerable country," says Pieter Huisman. "That is the key to understanding the Netherlands. Raising the dikes—it will never finish. You can only stop by leaving this country."

NOW AN altogether new peril threatens whatever time the Dutch have left in their water-bound land. The sea is rising. Nobody in the Netherlands scoffs at the idea of global warming; it would be madness to ignore it here in the lowlands.

I first heard about the changing glaciers from a friend in New Zealand. She told me in 1990 that a tribal chieftain in southern New Zealand warned her that the glaciers in his region were spreading. I couldn't make out what that meant at the time, but today it's no longer a mystery. Glaciers are losing height—melting—as the planet warms. It's a global occurrence. Glaciers in the Swiss Alps, which have lost half their size since 1900, are now losing between twenty-four and twenty-eight inches yearly. The Mendenhall Glacier in Alaska lost over half a mile in length in the final years of the twentieth century. The ice cover that reaches across the top of the globe is forty percent thinner than it was several decades ago. Around the edges of Antarctica, where the temperature of the air is rising faster than anywhere else on earth, ice shelves are dissolving dramatically. The Prince Gustav Ice Shelf, and the Wordie Shelf are almost gone and the Larsen Ice Shelf has collapsed.

Ice caps and sea ice are melting into the oceans because the world is getting hotter. The last decade of the twentieth century was the warmest on record. Since 1987, the planet has been warming at a heightened rate of 2° Celsius per century. In the past fifty years in the United States alone, the number of extreme heat stress days has increased by two days every decade. The "frost free" season in the

United States is eleven days longer than it was in 1950. Spring comes a week earlier. The incidence of heavy rain or downpours has increased by twenty percent. There are areas of the Atlantic Ocean that have warmed by a full degree since 1980.[2] That's fifty times faster than the warming that ended the last major ice age, the greatest sustained natural climate change in a hundred thousand years.

This is happening just as scientists predicted. The Intergovernmental Panel on Climate Change (IPCC), an independent UN-sponsored scientific body of 119 members from 32 countries, has stated in a series of reports released in 1995, 1997, and 2001, that in the next hundred years the planet will have warmed between 3° and 6° Celsius and that man has much to do with this change.

More frequent heat waves, droughts, storms, floods, changes in wind and rain patterns, and stronger El Niños will accompany warming.[3] Water supply will be profoundly altered. In a warming world, water will be sucked out of the soil at elevated evaporation rates. Less moisture in the soil will affect groundwater, rivers, crop yields, and even foundations of buildings. It will rain more in some places and less in others, so that rainfall won't correspond to reservoirs or irrigation systems in place, and precious water may dry up in established agricultural areas, such as southern California or southwest Australia. There will be dustbowls and new oases but we do not yet know where. Sea levels, already on the rise because of the thermal expansion of sea water and the addition to oceans of great amounts of evaporated water in the form of rain, may climb at a far faster rate than expected, a full meter or perhaps more.

Scientists at the Laboratory for Atmospheric and Space Physics in Boulder have discovered that as the atmosphere warms, trapping heat closer to the ground, the upper atmosphere receives less heat and is cooling fast, which could be global warming's most ominous component. These scientists fear that global warming in our complex system may trigger a flip in the other direction and usher in a new ice age. Since every part of our planet is interconnected—ice caps, landmass, heat from the sun, and ocean currents—scientists warn that warming could set off catastrophic cooling by melting ice in Greenland, say, or by making more high-latitude rainfall, which would dump more fresh water into the oceans. Readings of ice cores have shown that climate change happens abruptly, and therefore, scientists fear that an atmospheric inversion could happen in a hurry. "In human history," said the German magazine *Der Spiegel*, "far smaller temperature shifts have doomed kingdoms, set off wars, forced peoples into exile, and created new religions."

While virtually no one doubts the earth is warming, there are a few scientists who disagree on the extent of the causes and effects of such warming. The press tends to report global warming badly or infrequently, giving space to disagreements without making conclusions. It takes time to plow through the arguments. I've been keeping a file for twelve years, and while there have been many good articles on the subject, almost none of the magazines in which those articles appear are for sale in my small hometown. So, while many fear warming, they haven't caught up with what it means: It is altering the most basic systems on earth. "In a matter of decades, unless we change huge parts of our economies, we will live on a very different planet from the one we currently inhabit," says the environmental writer Bill McKibben. "There's the biggest story of our lifetimes by a wide margin, and yet polls show it finishes with the also-rans on any list of issues that people care deeply about."

"The greenhouse effect will not only influence the average temperature on earth and the sea level," reads *Rising Waters, Impacts of the Greenhouse Effect for the Netherlands,* a recent study published by the Dutch government, "but also wind, storms, rainfall, evaporation, and river discharge will change in a way we cannot predict at present."

The Netherlands is already sinking eight inches a century. With or without global warming, acres of dunes will disappear and the coastline will erode and become unsafe within a decade. In 1987, the Netherlands began investigating what would happen in case the sea rose still faster. At first, because changes from warming can be camouflaged by natural climactic variations, it was difficult to assess permanent deviations with confidence. But over the years, certain things became clear: more rain was falling between 35 and 70 degrees North latitude; sea levels had climbed as much as fifteen centimeters; high tides had become higher and low tides lower.

Holland's Rijkswaterstaat came to understand that if there were to be an accelerated rise in sea level, even "Delta Safe" would no longer be safe enough. As water temperature increases, so will algae. More rain, more storms, and more severe storm surges will put pressure on inland and coastal defenses. Saline seepage into inland water and the subsoil will increase. Drainage of low-lying areas will become more difficult, requiring more pumping. Bridges and dikes will have to be raised. The danger to those living in the delta and the heart of the Netherlands—Rotterdam, Den Haag, and Amsterdam—will be severely heightened.

In the lowlands, already struggling to cope with sinking land, water tables have dropped further and the land has fallen along with them, a process accelerated by superefficient modern pumps. In some areas the land dropped as much as two feet in the twentieth century. Subsidence has affected even land around Amsterdam, where every house on wooden pilings built after 1900 is at risk, since the reclaimed land outside the city center is not solid enough to hold them. As many as forty thousand houses will tilt and eventually collapse.

"We live in a sinking country with a rising sea," says Pieter Huisman. "So, there will come a moment when we have to leave this country." I gasped in disbelief, to hear a Dutchman admit such a possibility. "Yes," he grimaced. "I think it may take centuries, but there will come a moment when we have to leave this country."

IN 1953, an engineering student named Jacobus Van Dixhoorn was helping to clean up after the great flood, when he found in the muck, near an ancient graveyard, the skull of a thirteenth-century monk. "He made polders in Zeeland," Van Dixhoorn told me. "The monks who were cloistered here were dike builders. As a young engineer, I had found one of my predecessors."[4] Like the monks of Zeeland, Van Dixhoorn is a dike builder, a guardian of Holland's watery landscape, but his work is on a scale that his predecessors could not have imagined. The country he has cared for as an engineer has changed beyond recognition since 1200. About a quarter of the Netherlands—2,980 square miles out of a total surface of 12,300—has been wrested from the sea. Without the protection of the dikes, sixty percent of its land surface would be flooded every day. Yet it was the monks and the farmers they served who set the Netherlands on its strange and irreversible course.

The Dutch, long ago seeking refuge from the sea, built their homes on terps—high earthen mounds—linked by clay dikes. Even then, these descendants of German tribes were living on dangerous ground. "There lives a miserable people at the highest known level of the tides," wrote the Roman historian Pliny. "They have built their huts and live like sailors when their land is covered over and like the shipwrecked when the tides have gone out."

Roman generals Nero Drusus and Corbulo built the earliest canals in Holland in the first century of the Christian era. Irreversible intervention didn't really begin until a thousand years ago, as the growing population pushed into marshes and peat bogs along the North Sea. As the Dutch cultivated the lowlands, dug irrigation ditches, and built canals, slowly, surely, groundwater began to drop. As the ground water lowered, peat and clay compressed and the surface began to subside. At the same time, the Dutch were digging and burning great amounts of peat for fuel, causing the land to drop some more. As the landscape lowered, the sea began to wash over it. The Dutch built sea dikes to keep it out and interior dikes to protect their fields. They dug drainage ditches and dammed tidal inlets and creeks where the sea intruded into the countryside.[5] The Netherlands had embarked on a merciless struggle.

By the mid-fourteenth century, land in the northern, southern, and western parts of the Netherlands had subsided so greatly that gravity alone could no longer remove excess water. Netherlanders began to pump. Dutch millwrights borrowed the design of Mediterranean windmills but added rotating tops, which could catch winds from any direction, and put them to use, pumping water off the land. Row upon row of windmills were built to dry out sprawling parcels of ground. The newly reclaimed polders were surrounded with protective clay dikes strengthened by seaweed and wooden spikes.

Embankments were only as strong as their weakest points, and the task of maintaining them eventually grew too great for the monks. By the end of the twelfth century, a new system of caring for them was evolving. Each farmer was required to maintain his own portion of a dike—a solemn task, for if a poorly kept embankment broke, everyone in the vicinity went under. If a farmer couldn't manage, he was required to strike his spade into the ground and leave the property forever. Four upstream and three downstream neighbors then chose someone else to take over the property, a rigid protocol called the Law of the Spade.

Local societies elected representatives to oversee the care and building of larger waterworks that were beyond the means of local landowners. These elected "water guardians" formed regional and provincial *waterschappen,* or water boards, headed by a dike *reeve,* or dike lord. The water boards levied taxes, administered water matters, and doled out fines. Water boards in Delft and Leiden still own branding irons once used to deal with miscreants. At least one farmer was put to death for failing in his duty. "In this country the right of society is greater than the right of the individual," I was told by a flood-control engineer. "This comes from the water."

For centuries, the resolute Dutch have pushed against the sea—and the water always pushed back. When protection failed, the water could undo in a day work that had taken decades. In 1287, a tidal wave destroyed much of the north coast and tore open the mouth of a large inland sea, the Zuider Zee. Fifty thousand people died. On Saint Elizabeth's Day in 1421, the sea broke through both the coastal and the interior dikes, washed away twenty-four villages, and covered more than three hundred square miles of the Rhine Delta. During that flood, people stranded atop the Alblasserwaard Dike near Dordrecht are said to have seen a wooden cradle being washed along by torrents. In it, a cat jumped from side to side, steadying the cradle in the waves and thereby saving itself and the baby inside. The dike where the cradle came ashore is still called Kinderdijk, "the baby's dike."[6] It is also said that those survivors of the flood at Alblasserwaard stayed and became fathers to the best water-fighters in the world.

"The foe outside must be withstood with our common resources and our common might," wrote Andries Vierlingh, dikemaster to sixteenth-century king William the Silent. "For if you yield only slightly the sea will take all." Vierlingh's thoughts reflected those of his countrymen, a belief in a God-given right to fight the sea's "blind strength" with all of their own. The church, after all, had been the leader in land reclamation and the right—indeed the imperative—seemed in every way theirs. "The making of new land belongs to God alone," wrote Vierlingh. "For He gives to some people the wit and the strength to do it."[7]

The Dutch used the sea's "blind strength" against their enemies as well. To this day, the Netherlands holds an annual celebration of ending the eighty-year-long Spanish rule by use of its dikes and canals. The Sea Beggars, or Guex de Mer, mercenaries in the service of William the Silent, undertook a systematic breaching of river defenses to rout the Spanish by flooding Dutch cities. Their great triumph came in 1574 when, after breaking through a succession of dikes, they sailed from lake to lake in flat-bottomed boats and, with the help of a wild September storm, broke the five-month Spanish siege of Leiden.

In the sixteenth and seventeenth centuries, Holland's merchant fleets sailed the world and returned heavy with the spoils of trade. At one splendid moment in time there were more Dutch merchant ships than those of all other European nations put together. Wealthy merchants invested in new, deeper canals and helped finance the reclamation of hundreds of thousands of acres of land. In 1612, a Dutch drainage expert, Jan Adriaanszoon Leeghwater—his name means "empty water"—used forty-three windmills north of Amsterdam to dry out an

18,000-acre expanse known as Beemster Lake. Leeghwater believed that draining Holland's inland lakes was among the "most vital, most profitable, and most sacred tasks, which faced the whole of Holland," and before he was through he had turned 80 square miles of inland water into good farmland. Leeghwater wanted to employ 160 windmills to drain Haarlem Lake, 150 miles square, but the fulfillment of this dream had to wait until the end of the eighteenth century and the advent of steam-driven pumps, powerful enough to drive water over high dikes. In the mid-nineteenth century, long after Leeghwater's death, Haarlem Lake was emptied by just three steam-powered pumping stations.[8]

In 1916, after a storm loosed particularly terrible floods around the Zuider Zee—a thirteen-hundred-square-mile estuary that opened into the North Sea—the Dutch resolved to cut it off from the marauding sea by building a barrier dam across its mouth. Under the leadership of another legendary engineer, Cornelius Lely, Netherlanders first laid willow mattresses on the seabed to protect the floor from scouring, and then employed a battalion of steam-powered dredgers and draglines to bring up clay from the bottom to build the dike. They worked from both sides toward the center, and as the work progressed and the channel narrowed, the volume and intensity of water pouring through the opening became more and more powerful. In spite of the current, they managed to close the twenty-mile-long dike in 1932, turning the unpredictable Zuider Zee into placid, freshwater Lake IJssel. Fifteen hundred square kilometers of land were then reclaimed out of its waters. Many of the Dutch feel that the closing of the Zuider Zee was their most spectacular achievement, since their only tools were steam and sturdy men.

NOWHERE IN all of precarious Holland is the battle against the sea fought more intensely than along the exposed edges of low-lying land in the southern Delta, which is pressed by water, front to back. Three rivers flow out of the interior of the country into the Delta—the Rhine, the Scheldt, and the Maas—branching into a network of tributaries, tidal channels, and estuaries that empty into the North Sea. The land of the Delta, rich with alluvium carried by the rivers, is heavily farmed. To the north is Rotterdam. Southward lie South Holland and Zeeland provinces, low and lower, respectively. The Delta has been in upheaval for such a long time, as the sea moved in or the Dutch reclaimed land, that generations of maps bear little resemblance to one another. On very old maps there are myriad tiny spits of earth, little more than terps. As reclamation progresses, the maps shift gradually to show more earth than water.

In 1953, as Pieter Huisman had told me, calamity struck the Delta. Full spring tides and a nasty northwesterly storm created a tidal wave that devastated the lowland, breaching 400 dikes and inundating the land almost as far north as Rotterdam. Telephones were washed out so the rest of the country remained unaware of the scope of the disaster. When it was all over, a shocked nation learned that 1,865 of its citizens had died. In Zeeland, the hardest hit province, some villages lost ten percent of the population. It was a defining moment for the Netherlands. Within one week, a Delta Commission was formed and within three years the Rijkswaterstaat had come up with a plan to make sure that the people of the Delta would never again have to fear such a catastrophe.

Jacobus Van Dixhoorn—known to his friends as "Co"—was studying engineering then, in Delft. The city wasn't flooded, even though it is below sea level, and Van Dixhoorn's chief memory is of streets filled with roof tiles blown down by powerful winds. He quickly made his way to the flooded area to help clean up and there found the "death's head" of his dike-building predecessor. Soon thereafter, he joined the Rijkswaterstaat. Eventually he became its general director.

Van Dixhoorn is tall and rugged-looking, a descendant of Calvinists and watermen. Like so many of his fellow Dutchmen, he is exceedingly modest, with a profound distaste for hubris. He welcomed me to his unassuming home in a green area of Den Haag, its garden pushed up against a tree-lined canal, its only vanity extravagant bowls of peonies and roses placed around the house. I asked him why he had become an engineer. "Probably it was somewhere in the blood," he said quietly, a warm smile softening his craggy features as he talked about his work.

"*Waterstaat* means the 'state of the waters,'" he said, "which means the physical existence of this country. First, you have to defend half the country against the sea. You are situated behind dikes in a low area, so you must pump. The part of Holland that is above sea level suffers a shortage of water, and that area is covered by agriculture, so fertilizers and manure poison the ground water. If you want to do something with your waste, you can put it on land, but it will soon arrive in your drainage water. It all comes back to water."

The Rijkswaterstaat also handles transportation routes. "Roads are not natural to a lowland country, especially not in a delta, so you have to spend enormous amounts of money to build land connections," Van Dixhoorn explains. "If you cross a river with a bridge, you stop seagoing traffic, so you need a tunnel, which in this low country is not an easy thing. Before you know it, it has turned into a sewer. So the environment, physical planning, and what we call *waterstaat* are closely connected."

With water pushing under, over, and through the cat's cradle of ditches, dikes, canals, and rivers that comprise Holland, the need for a single administrative body to oversee all the affairs of water in the country is incontrovertible. Since its creation in 1798, the Rijkswaterstaat has become the most efficient national water agency in the world, driven by the Dutch tradition of simultaneously defending and inventing the country. Van Dixhoorn is just another in a long line of stouthearted generals in the ongoing war for ascendancy over the sea and rivers.[9] But it was on his watch that one of the boldest, most breathtakingly innovative engineering projects in history was launched: the building of the Dutch Delta Works and its centerpiece, the Oosterschelde Storm Barrier.

AFTER THE 1953 FLOOD, the agency committed a hefty portion of the country's resources to ensuring that such a thing would never happen again. Within weeks, the Rijkswaterstaat decided to close most of the deep channels in the Delta. Parliament approved the plan in 1958, and the nation geared up for a twenty-five-year offensive. The Rijkswaterstaat would make it up as they went, inventing fantastic new sea barriers and mighty sea vessels to meet the specific demands of each body of water, of each spit of land. The geographical challenges were daunting. Delta storms are fearsome—even everyday winds and currents here are formidable—so the stamina and ingenuity of engineers was to be sorely tried. While engineers were building to keep out the North Sea, they had to also allow the movement of the rivers flowing west.

Rijkswaterstaat engineers began near Rotterdam. The long, broad stretches of Holland's main port—the country's "front door"—had to be kept open, while the Hollandse IJssel, a smaller river to the east, could be protected from storm surges by a barrier that engineers called "the latch on the front door." Next the Dutch built two caisson dams in the southern Delta, the Zandcreek and Veerse Gat, in 1960 and 1961. Caissons for underwater dam-building are watertight hollow concrete structures, first set in place, then filled with sand or stones. There are seven caissons in Veerse Gat, each seven stories tall, each set on a sill of stone, filled with sand, and a dike laid over them.

In 1965, Dutch builders finished a four-mile-long dam at Graevelingen, and closed part of the channel by dumping 170,000 tons of stone from gondolas suspended from a steel cableway.[10] Several ordinary dams went up too and finally a three-mile-long dam at Haringvliet, with gates on both river and sea sides.

But the great challenge still lay ahead: closing the particularly broad, especially deep eastern branch of the Scheldt River. Three long tidal channels in the deepest part of the estuary, ranging from under fifty to over a hundred feet deep, would have to be blocked by some sort of massive physical structure. Until now human prosperity and safety has been virtually the only concern in the Netherlands' water wars, but by the time a strategy began to evolve for what came to be called the Oosterschelde Dam, ecological concerns had become part of the planning. Environmental care would be especially critical in the Eastern Scheldt.

Behind the mouth of the channels lay a tidal inlet, filled with plant and animal life—seventy-five kinds of fish, twenty-five bird species, nearly a hundred and fifty species of algae and water plants, and almost two hundred of plankton. Building a dam across the Scheldt would cut off the saltwater tides that replenish the inlet every day, wiping out both the natural system and a bustling mussel and oyster industry. After prolonged, often heated discussion, Dutch politicians and the Rijkswaterstaat abandoned the idea of a solid barrier dam and opted for a storm-surge barrier, a far more complicated and expensive structure.

On a sunny day, if you are standing in the heart of the Oosterschelde Storm-Surge Barrier, even the benign pounding of wind and tide makes your ears ring. High tide moved in while I walked on the barrier, and its ferocity made me think hard about the wisdom of challenging the sea. Even on a relatively calm day, the sea is terrifying. The thousand-ton concrete beams at the top of each gate seemed to tremble under the battering of the elements, but the workers manning the structure were unconcerned. They knew what they were standing on. Willem Zwigtman, a handsome young Dutch construction engineer who helped build the barrier, pointed out that while the beams may feel a bit fluttery, the base piers are entirely unmoved. "The pier is like solid rock," he says, in awe even after years of working at the site. "There is no movement at all. If the piers move, then the gates won't go down. This is why everything in the barrier has to be so precise, so exact."

There are sixty-two solid steel gates in the storm-surge barrier, hanging above the sea between sixty-five colossal piers. Each pier is the size of a twelve-story building and is embedded in a vast artificial mattress in the seabed, anchored by stone and holding gates as wide as half of a New York City block. Zwigtman is dead right. The thing is astonishing. How do you position monstrously sized, prefabricated parts and assemble them in the open sea to an accuracy of one or two centimeters, so that nothing moves or shifts or compresses, yet with strength enough to withstand the jet force of the tides?

Zwigtman remembers initial opposition to an open barrier. "In the beginning, people complained that it was too expensive," he comments. "Some said, 'Just heighten the dikes.' I knew engineers that worked on other Delta dams that said, 'You cannot do this. Just close it with a solid dam. You cannot prefabricate a thing like this. We know the power of the sea and its strength, and we do not believe this will work.'" But the Rijkswaterstaat had learned a lot by building the earlier dams in the Delta and knew they would learn more with this one. It had been many years since a Vierlingh or a Leeghwater achieved fame from their works, and the names of the many engineers who designed and executed the components of the Oosterschelde Storm-Surge Barrier lie camouflaged between the pages of articles in scholarly journals. The nation would take the credit.[11]

Engineers first dredged several islands out of the shallowest parts of the estuary of the Eastern Scheldt, which would first be used as fabrication sites and eventually become part of the barrier. Then the dredgers gouged a deep trench out of the clay seabed and laid polypropylene mats on each side of the furrow to stabilize the channel floor and form a solid bed. Although porous, the sand brought into the center trench to replace clay could be made foundation-worthy by compacting. For this, Dutch engineers invented a floating pontoon barge, the *Mytilus,* which was outfitted with a fifty-foot-high rig supporting four outsize vibrating needles to shake down the bottom of the trench.[12] It took three years just to stabilize the trenches in the tidal channels.

The piers comprising the base of the barrier required a perfectly solid, level floor, which the engineers had to create by laying mattresses made of polypropylene, reinforced with steel cable. The mattresses were 656 feet long and 138 feet wide and contained layers of sand, gravel, and stone. They would render the subsoil impervious to erosion, prevent weakening due to shifting water pressure, and ensure evenness to a tolerance of plus or minus four inches. To transport and precisely place these rock-filled mattresses, the Dutch specially equipped a square barge, the *Cardium.* She had twelve suction nozzles attached to dredge pumps at her stern, which first vacuumed loose debris from the bottom of the trench. On her bow was mounted a great spool-like drum onto which the mattresses were rolled. Sophisticated positioning equipment in the control tower compensated for wind, current, and wave movement while computers controlled a system of winches that delicately dredged, vacuumed, compacted, and, as the *Cardium* moved backward, laid the mattresses.

If they were not on the seabed by the time the tide started moving out, the force of the water would destroy them. "You had about a half hour to roll two hundred meters of mattress out on the sea floor," Willem Zwigtman told me. "You had to put it down in just one move, because you couldn't pick it up again. Once, a mattress fell off the drum because somebody pushed the wrong button. The whole thing was ruined, but I had an extra one ready to go."

Sixty-five mattresses, each weighing 5,500 tons, were laid precisely in place. Sixty-five smaller ones were then placed on top. Wherever extra leveling was needed, smaller synthetic sheets topped off the first two mattresses. When the job was done, the tidal channels had been steadied by five million square meters of mattresses. Its trench beds were as flat as most football fields.

Meanwhile, the sixty-five concrete piers that would make up the backbone of the structure were being manufactured in construction docks on a specially created polder, fifty feet below sea level. The piers were thick at the base, 165 by 82 feet, but narrow at the top, to minimize the obstruction to water pushing through the barrier. Once in place, the hollow piers would be filled with sand. Their top half consisted of configurations of shelves and notches to accommodate gates, beams, and a roadway. Each pier was cast in one uninterrupted pouring that lasted more than 36 hours. Since it took a year and a half to build each pier, work started on a new one every eight days.

Moving each twelve-story monolith fourteen miles from its construction site to its trench in the sea was another challenge the Dutch had learned to meet. When the piers in a construction dock were finished, the dikes were broken and the dock flooded to sea level. The *Ostrea*, a

U-shaped lifting vessel so big that she dwarfed the piers, floated into the dock, clamped hydraulically operated claws suspended from hoisting frames to the sides of the twenty-thousand-ton pier, lifted it, and moved out of the dock toward the barrier.

Since the draught of the *Ostrea* was about 36 feet, she had to find deep water. In two of the channels, this meant sailing up into the Scheldt and back again. "To place your pier, you had to sail far to the east and come back along the coast," Willem Zwigtman told me. "So you were busy a full day with your pier, and you knew if you were not at a certain point at a certain time in the Eastern Scheldt, you were in trouble. If I missed one tidal movement, which is six hours, well, six hours delay on a project of this scale costs a lot of money. But it all went very well."

Zwigtman and his men had to constantly jockey with the tides. "We have a current speed of about five meters per second in the channels, which is tremendously fast," Zwigtman says. "The current was so strong that you couldn't hold even these huge structures. So you could only work in a period that was more or less tidal free."

A positioning vessel, the *Macoma,* waited for the *Ostrea* over the foundation bed, attached to anchor piers built into the sea floor. "We then connected the *Ostrea* to the *Macoma,*" says Zwigtman. "Since you only had half an hour to lower your pier, you couldn't sail there and then start to figure it out. We used a positioning barge." When channel velocity dropped below a meter a second, the lowering began. A small underwater inspection vessel with five cameras mounted on it darted around the sinking pier. When the pier was in place, the *Ostrea* loosed its claws, detached from the *Macoma,* and returned to the dock to pick up another pier. Each monolith was placed in under an hour, in a position exact within tenths of an inch.

Next, bags of gravel and rubble were dumped around the base of the piers and the foundation grouted by forcing sand-cement mortar through a network of pipes into the base cavity between the bottom of the pier and the foundation bed. The hollow lower vaults of the pier were filled with sand. A deep sill of graded stone was placed along the piers and compacted. Finally a layer of six- and ten-ton stones—too heavy to be carried away from the barrier by even the most crushing currents—was placed on top. It took two years and five million tons of stone to build the sills.

With the sills complete, workers moved in on an eighteen-hundred-ton floating crane, the world's strongest, to put the superstructure in place: road-bridge box girders; capping units; the upper beams and a hydraulic cylinder system to open and close the sixty-two sliding gates—each weighing between four hundred and six hundred tons. The gates have to be lowered slowly—it takes an hour for the largest one to close—since if it is done too quickly, the movement of the gate itself will cause a translation wave in the Scheldt. Fifty staff members work at the barrier in a centralized-control building, but it can function without anyone at all. "The whole thing is computerized," says Willem. "Which means that even if nobody is in the control room, and the sea level rises to three meters, the barrier goes down."[13]

"It's nice to think that all these pieces—the mattresses, the piers, the gates—were made elsewhere and fit together within centimeters," Zwigtman says. "Nice" is Dutch understatement. Here in the Eastern Scheldt stands a piece of engineering so audacious, so spectacularly effective that it takes your breath away. You cannot walk across this colossal structure without loving Dutch ingenuity—and Dutch stubbornness.

I asked Willem what was done with the sea vessels used to build the barrier after it was finished. "There were some wild ideas in the beginning," he told me. "They thought when this was finished, maybe they could use the *Ostrea* to lift the *Titanic*. But it didn't work. You can't do anything with these ships. Nowhere else in the world are they going to build anything like this. It requires so much political will to spend so much money for ecological reasons that I don't think any countries will spend the extra billions."

"What finally became of them?" I asked.

"They were sold as scrap metal."

ON OCTOBER 4, 1986, Rijkswaterstaat General Director Jacobus Van Dixhoorn watched as Queen Beatrix of the Netherlands formally opened the Oosterschelde Storm-Surge Barrier. Bands played, balloons rose, champagne flowed, and fireboats and tugs filled the air with spraying water and the blare of foghorns. It was a great day for the Dutch.

The storm-surge barrier took ten years to complete. It cost $3.2 billion out of the complete Delta Works bill of around $5 billion. The open barrier in the Eastern Scheldt cost another $800 million. The dams in the Delta have a flawless record. Zeeland, a province whose brave motto "I struggle and rise up" is still emblazoned on its crest, lies cozy and safe behind the Delta Works. Millions of people come here for recreation every year, to boat, fish, walk the dunes, and circle wildlife-filled estuaries.[14]

Willem Zwigtman offered to drive me around Zeeland's soft green landscape. Nearly everything about the place seemed ordinary enough. The fields were fat with orderly squares of wheat and alfalfa. Rows of manicured trees lined the roads. There were churches, taverns, factories, and banks. Men have settled in to stay. Yet the horizon seemed to be in the wrong place. Boats sailed by over my head. I was perpetually uneasy, aware of the North Sea above me and of the indistinct blending from green to blue in the distance. The Zeelanders seemed perfectly secure in their sweet villages of brick streets and small homes with tidy kitchen gardens, but I don't know that I could ever feel at home in their submerged landscape.

Zwigtman took me to one of those charming houses to meet his pretty, blond American wife. We talked about Frans Hals, blues music, weekend trips to Antwerp, and how much the couple likes living in the Delta. The Zwigtmans believe its people are special. "When you go to coastal areas, people are more open. They are not as conservative as in areas where they don't see the sea."

Zwigtman is typically Dutch—not boastful, but quietly pleased by Holland's achievements in the Delta, by the understanding of the sea its people have gained over centuries of struggle and coexistence. "Although a lot of the new dams and machines were thought up by young civil engineers," he said, "on the Oosterschelde project we were often guided by engineers who had been shaped by day-to-day life, who had more feeling about what is happening outside than guys who know it from books and calculations. It's not just theoretical knowledge—we've grown up with the sea. We understand what high tide and low tide means."

And now, armed with their understanding and fully comprehending their own vulnerability, the Dutch have begun to call a halt to their aggression against the sea. A 150,000-acre polder in IJsselmeer, in the works since 1932, has been abandoned. The government believes that the cost of defending such a polder is too high.

Even more astonishing, the Dutch have started to voluntarily surrender land. In a carefully contrived plan begun three years after the Delta Works were completed, the Netherlands has agreed to give up a full tenth of its farmland. Six hundred thousand acres of dry land are now becoming lakes, wetlands, and floodplain forests. In Flevoland, waters have been allowed to flow back over 120 hectares to create the Duursde Waarden wildlife reserve. The reclamation has brought cormorants, spoonbills, and marsh harriers back to a landscape that had lost half of its birds in the previous fifty years. Thousands of farmers are abandoning agriculture to become paid stewards of their own naturalized land. Just south of Amsterdam, near Mijdrecht, where farmers sometimes have had to pump 16 hours a day, 9,000 acres are being flooded.[15] The Netherlands has graciously conceded some land to the water.

Holland is safe for the moment. Its watermen, however, can never relax their vigilance. "It requires an enormous amount of money to keep this country dry, to pump, to maintain the dikes and warning systems," Van Dixhoorn reminded me. "That makes living and working and industrializing and farming in Holland more expensive than anywhere else." Van Dixhoorn, like most of his countrymen, never doubts that the expense is necessary. "Since half the country is situated below sea level, you have to defend half the country against the sea," he says. "You have to pump. Day after day, you have to pump."

When the Dutch first stalwartly climbed onto their terps, they began a profound legacy of work for their descendants—endless, mortal combat with the sea. Sixteen hundred years later, engineers have proven that they can build improbable and magnificent tools to protect their people's handholds on water-lapped shores. But even as coastal inhabitants understand that if they wish to stay where they are, they cannot let their guard down for a moment, the costs of protection rise. The Rijkswaterstaat already spends $400 million a year to pump and repair dikes. With such astronomical costs and the seas rising, an obvious question to ask is if people living in such peril should stay or go.

Johan Van Veen, called the father of the Delta Plan, raises the question of economy in his book *Dike, Dredge, Reclaim*. As an example, he asked whether the money spent on saving the delta island of Schouwen—about five times its value—was justifiable. "Did you never think about abandoning that island?" he writes. "How many times after each flood in the past have you reconquered and remade your lowlands?" But Van Veen is merely posing the question. He goes on to say that he has never heard the value of reclamation challenged by a Dutchman and never expects to hear it. "Necessary wars," Van Veen writes, "the Dutch struggle against the sea is such a war—ask for higher reasoning than mere short-term economy can provide. . . . Economy is the servant, not the master of life. The right to exist comes first . . . yes, even before economy. The whole of the Delta Works can be had for one year's army budget, a mere trifle in the state economy of centuries."

Farm products alone, grown on the Netherlands' artificial lands, are worth billions to the Dutch economy. And so the Rijkswaterstaat continues to fiercely defend what it deems essential. Holland has recently completed another immense, one-of-a-kind, technological magic trick. Like all the Dutch superstunts, it looks simple, having been nearly impossible to accomplish. They have closed the doors of Rotterdam Harbor, the world's largest, busiest port, with a set of gigantic gates, each longer than the Eiffel Tower is tall. Hinged on the world's largest ball-and-socket joints, aluminum bronze spheres each as big as a house, the gates float out of the way of ship traffic when not in use but quickly and efficiently close against North Sea storms. Rotterdam's protection cost a relatively modest $450 million.

Although there seems to be little questioning in the Netherlands of the Rijkswaterstaat—they have, after all, kept the country afloat for centuries—at least one Dutchman calls into account everything about his countrymen's phlegmatic acceptance of their existence. For another view of the relationship between man and water in the Netherlands, I went to see Dutch filmmaker Louis Van Gasteren in a high, narrow house on a canal in the center of Amsterdam. Unlike the unpretentious houses of the watermen, it is filled with art and artifice; good pictures, elegant objects.

Van Gasteren is an artist but looks like a burgher; tall, beer-bellied, and red-faced under a great shock of wild white hair. "Water is not fun," he announced provocatively on my arrival. "There are no fountains in this country. With water one should not play. It is dead serious." Ironic, and impatient with the failure of his countrymen to add whimsy to water, Van Gasteren's views nonetheless run a complex gamut of pride, approbation, and a grasp of Holland's strange legacy. He believes that out of the struggle to control water has come a uniquely independent country, but one that has pumped itself into a conundrum.

"Freedom is vital for cooperation and both are necessary to keep our heads above water," he tells me. "I believe that this has influenced the Dutch character. And that makes this country very, very special." Van Gasteren reminds me that the Dutch were the first to establish a recognized measure of elevation—sea level. "To agree on this is to understand the globe, how high the climber is on a mountainside, how deep the workers are in a mine pit," he says. "For this you need imagination. But this same imagination has provided us with a simple explanation of this country—the little Dutch boy with his finger in the dike. The answer is not the finger in the dike. The answer is a highly technological operation. That is why things function in this country!"

Louis Van Gasteren, after an intense consideration of Holland's centuries of pumping, has made a documentary film, *A Matter of Level,* which suggests that the Dutch determination to control their water may be a mass neurosis for which the only diagnosis is emigration. Van Gasteren gave me a private screening of his offbeat film. "There was a time that man saw a flood as a punishment from God," drones the narrator. "The Dutch are a practical people. They created a whole separate culture, a water culture. The North couldn't exist without its watermen." Here Van Gasteren's Dutch pride begins to stumble over the complexities of the conflict between technology and a people who don't comprehend its costs. His face reddens, the wineglass in his hand rises then lowers. He interrupts the film. "This country is under water," he shouts. "So you will not see a damn thing of all this technology. Fifty million people do not know anything about their existence, except that they have dry feet."

At the heart of Van Gasteren's argument about Holland and water is a deep ambivalence about the Dutch struggle. "We have labored to shape this country," he groans. "The Dutch have nature in their hands. Every square meter is the result of human effort. There is no untouched nature and we are all responsible. This is a non-natural land!"

Van Gasteren's film finally runs to its finish:

"Pumping for all eternity. That's land reclamation. Pumping forever! It's a matter of level.

Question: Did no one perceive this great ordeal a thousand years ago?
Answer: Man is not equipped for that."

"Frisia non cantat" is an ancient Latin saying that means Friesland—the

lowlands—does not sing. Perhaps not, perhaps the seriousness of the Dutch struggle against the sea has helped to keep levity as well as fountains to a minimum. But it's hard not to admire Dutch courage and technological imagination. In Andries Vierlingh's sixteenth-century book on water engineering, *Tractaet van Dyckagie,* there is a story uncomfortably like that of the little Dutch boy that Louis Van Gasteren so detests. During some maintenance work, the sea ruptured a dike and poured into a polder. "But immediately a stalwart Frisian threw himself into that hole and the other people came quickly to help and so this Frisian saved the land." The Dutch war against the sea, always a matter of individual hardiness and collective technical achievement, started long ago and continues.

"God created the World," said Voltaire. "But the Dutch created Holland." And so they did. Pieter Huisman suggests that appreciating the duration and ferocity of its struggle puts the Dutch effort in perspective. "History gives distance," he says. "We have been dealing with water intervention in this country for a thousand years, but when you put what we are doing on the long-time scale of history, it is not much. But it is irreversible.

"As an engineer you specialize in a science but live in a society," he continues. "The question is, what to do with other areas in the world—for example, the Mississippi Delta—migration or not? In the Po Delta in Italy, they are creating the same problems. And you have to give thought to the huge population of Bangladesh, for which diking and pumping is no answer. But *this* is a prosperous country," he says emphatically. "One of the most prosperous countries in Europe. If you want to stop pumping you will have to leave, but I think we still have some time to live in this country."

The Netherlands is not alone in its fear of a rising sea. Other places around the world share the perils and some of the technology as well. There is a partially completed set of dams and barrier gates just outside of Saint Petersburg in Russia, and there are artificial islands in Kobe Bay in Japan. East of London, the stunning Thames Barrier Gates were built as a response to surge tides in the river driven by winds off the North Sea, especially a particularly savage flood in 1953 that drowned 300 people. When closed, the gates present a 1,706-foot wall of steel against the encroaching water. The barrier has proved its worth: the gates have been closed twenty times since they were inaugurated in 1984. But the place whose problems most closely resemble those of Holland's Eastern Scheldt is Venice.

"ELEMENT OPPOSES ELEMENT," wrote an observer of the Lagoon of Venice in 1718. In this place, where man has been meddling for many centuries, it has never been truer. Venice has been sinking into the Adriatic for a thousand years but is now threatened more seriously than ever before. In the low-lying streets east of the Piazza San Marco, duckboards lie in the streets so that pedestrians can walk on them during floods. Lately no one bothers to push them to the sides since everyone knows the water will arrive all too soon. Rising waters have already damaged virtually every building in Venice. The Piazza San Marco, once inundated by high tides eight times a year, is now flooded on an average of one day in three. If the sea rises a foot in the next century, as even conservative scientists predict, San Marco and its surroundings will be flooded on all but a few days a year.

The gentle subsidence of the soil under Venice's marble palaces and churches began almost as soon as they were built. The mud of the small low-lying islands into which the people of the Veneto rammed millions of stabilizing saplings was soft and silty, and its compaction has occurred naturally. As marble and stone and brick were added over centuries, the city sank, imperceptibly yet relentlessly. When more water was pumped out of the ground to meet the demands of twentieth-century hotels and modern industry, the land dropped still faster.

The slow sinking of the city is countered by the *acque alte*—the unusually high tides that come from the Adriatic Sea, intensified by winds blowing east out of Syria and south from the Hungarian steppes—which push persistently into the Lagoon of Venice. They wash over the low parts of the neighboring island cities, Burano, Torcello, Murano, as well as the villages sitting on the breakwaters that enclose and protect the lagoon: Palestrina, Malamocco, and the Lido.

There is more to the anguish of Venice than the Adriatic rushing through its streets. The rivers that once fed and flushed the bay above Venice were diverted centuries ago to prevent silting and to ensure easy navigation, and over time the lagoon's shallow waters have become still. The thousand tiny little islands, or *barene*, of the Lagoon of Venice have virtually disappeared, deprived of the river's silt and eroded by wind and the incessant wakes of boat traffic. The seabed has flattened, its natural life depleted. The intensively farmed lands of the Po Valley carry insecticides and fertilizers to lagoon waters, and the petrochemical industry at Marghera adds heavy metals, so that marshes and waterways suffocate with algae.[16] The lagoon is dying from a cycle of deprivation and decay.

Water has always been the strength and prosperity of Venice, and its leaders once understood it well. They created the Magistracy of the Waters, a powerful agency of the Venetian Republic that legislated water use and protection across centuries. Every fisherman, every man that wished to sink a piling or dig a channel had to deal with the Magistracy, which took the broadest possible view of its charge. After a time, Venice's understanding of the delicate balance of its watery world began to fade. The Magistracy drifted into the role of plumbers, and Venice and its surrounding islands drifted into trouble. Without the guiding hand of an agency whose premier concern was for the water itself, the lagoon began to die.

In November 1966, the worst *acqua alta* in history brought six feet of water into the lagoon, and gave rise to an international frenzy about how to save Venice. Since then, controversy has raged over whether to save the buildings or the city itself. Improbable solutions have been proposed, such as creating inflatable rubber-sausage barriers or moving the entire industrial port of Marghera. Every year glossy magazines run photographs of "Save Venice" balls given by socialites in pricey costumes. The cry for Venice can be heard too in the frustrated moans of its citizens, in the sour, hopeless articles in the press and the self-serving noises of politicians. In the end, if Venice can be saved—and there are those who are determined to do so—it will be by a familiar cast of characters: engineers and politicians.

The elegant marble offices of the Consorzio Venezia Nuova, a consortium of twenty-one Italian companies charged by the government with saving Venice, sit just behind a large courtyard in the Dorsoduro, centered around an ancient cistern, one of many in which Venetians once collected rainwater for drinking. Upstairs in the Consorzio's ornate, high-windowed palazzo, I met its technical chief, Deputy Director Maurizio Gentilomo, a block of a man with a huge smile, elegant Italian suit, and classic Italian charm. Gentilomo was delighted when he discovered that I was familiar with the Tarbela Dam in Pakistan, since he once worked for Impregilo, the Italian engineering firm that helped build it 26 years ago. "A portion of my heart is in Tarbela," he sighs. Gentilomo is an engineer—a builder of dams, bridges, and ports—and his mission is to rescue Venice from the sea and the lagoon from the intrusions of men.

"The problem of Venice," Gentilomo explains passionately, "is that ground floors of the buildings are frequently flooded during the high tides. These tides occur twice a day depending on the astronomical factors—the attraction of the moon—and the meteorological factors, which are variable without warning."

The crescent-shaped Lagoon of Venice runs about thirty miles north to south and is about six miles wide. Venice itself sits in the central lagoon, a city of roughly seventy-five thousand year-round residents. "The natural sink of Venice," Gentilomo continues, "is around one millimeter per year, which means that in a thousand years the city lowers one meter, as it already has, since Venice is over a thousand years old. That is ten centimeters each century. Simultaneously, there is an increase of the mean level of the sea."

In 1973, the Italian Parliament created a special act that defined the

Lagoon of Venice as of "preeminent national interest." Under this legislation, a number of projects were imagined, none of which passed
hydraulic, economic, or environmental muster. And so, eleven years
later, in another special act, the Consorzio Venezia Nuova was given two
tasks: the job of managing the research, design, and implementation of
defending Venice and the other towns and villages in the lagoon against
flooding, and of finding a way to salvage the complex hydrogeological
system of the lagoon itself. In 1992, the Italian Parliament allocated $1.5
billion for its work.

The engineers have already made important progress in stabilizing the
lagoon's environment. They have fortified the eighteenth-century
murazzi—the sea defenses along the outer protective islands—by raising
waterfront walls and reinforcing breakwaters, groins, jetties, and lagoon
mouths with rock and sand. This was an enormous task, since it involved
building up both sides of the long narrow islands that stretch thirty miles
along the lagoon's edge. They have built new flood protection and
drainage systems around Sottomarina, Pellestrina, San Pietro in Volta,
Malamocco, and Treporti to stop water seepage. And in the lagoon
itself, new sewage treatment facilities will soon cut the inflow of industrial pollutants in half. If pipelines are built to carry oil, tanker traffic
will be removed from the inland waters, where an oil spill would be disastrous.[17] The engineers have also planted tracts of pollution-absorbing
vegetation to eat up chemical runoff from the farms.

The Consorzio, in a dazzling restoration of the lagoon's ecosystem,
has reconstructed wetlands and reworked lagoon hydrodynamics to
stimulate water circulation.[18] Engineers dredged deeper channels to get
the water moving. Using dredged sediment, they have reconstructed
shoals and built up myriad *barene,* fortifying them with logs and fast-
growing plant life. On the lagoon floor they introduced oxygen-
producing aquatic vegetation to stabilize erosion and make havens for
microorganisms, mollusks, and fish. This water, sluggish for decades,
has begun to move again. Vegetation and marshes are stabilizing, and
water birds have returned to the lagoon.

But the Consorzio had not yet solved the enigma of the sinking city, a dilemma only partially similar to that of the Dutch Delta. The outer islands that protect Venice were formed by thousands of years of silt surging into the lagoon and settling at its rim. In between these islands run three channels through which tides enter the lagoon to flush, clean, and refresh it. If the cleansing exchange of water with the sea were the only concern, a barrier similar to the Oosterschelde Storm-Surge Barrier could be constructed. But because the industrial port of Marghera is located at the top of the lagoon, large ships continuously pass through the channels, and five hundred fishing boats sail out of the lagoon into the Adriatic and back again each day. A solid barrier gate wouldn't work.

Instead, the Consorzio has come up with a complicated and unprecedented solution, a set of eighty flap gates to be built across the inlets to the sea. Sixty feet wide and hydraulically operated, these steel caissons will lie flat on the channel floor when not in use, allowing boats and tidal waters to move freely above them. But when the tides run unusually high, they will be pumped up, rising upright to close out the sea entirely.[19]

"The gates are moved by Archimedes buoyancy," explains Gentilomo. "Steel boxes, hinged along one side, are filled with water when sleeping, so the weight maintains them horizontally on the bottom of the channel. When we need to operate the gates, we will expire the water inside the box by means of compressed air through the hinges. The element moving the gate is always fluid, no rams, no chains, no levers. When the sea and the lagoon are again in equilibrium you can rotate the gates down by filling the box with water."

Between 1988 and 1992, the Consorzio built and tested a prototype gate called MOSE—an Italian acronym for "electronuclear experimental module" and also a sly reference to Moses and the parting of the seas.[20] Work on the actual gates was scheduled to begin once an environmental impact assessment was finished, but for almost ten years nothing happened. Engineers had established that the gates would not interfere with the tides moving through the channels, and the Consorzio had already demonstrated that they understood the hydrodynamics and the ecosystem. They had a plan. But nothing in Venice is ever simple.

Early in 1998, an international panel of experts recommended that the gates be built. In December 1998, the Italian government voted against them. The plan's detractors claimed that the gates would harm the environment, that so much cement on the seabed would be damaging, and that because thirty-five years have passed since the disastrous floods, it would not happen again.[21] Alternative plans were ventured, but these schemes—such as raising the city itself—often seemed so obviously flimsy that it's hard to believe that the protestors had investigated the facts. "They have forgotten the disaster of thirty-five years ago," says Gentilomo. "They think that small floods are acceptable—that if you buy rubber boots, which are cheap, you can afford floods.

"Byzantine!" he sniffs. "Simple problems are made complicated. They are *against* in principle." Gentilomo tells me that in this most man-made of environments, many Venetians bizarrely oppose anything they consider unnatural, citing Venice's scandalous lack of a sewage system. "They don't seem to realize that the Basilica San Marco is a fully artificial object," he protested.[22] As the political wrangling continues, Venice continues to sink and the sea to rise. In the twentieth century alone, the land dropped five inches and the sea rose by about four inches, a net loss of nine. "Time," Gentilomo says to me softly, "is of the essence in Venice."

IN 2000, an independent panel of experts declared that MOSE was the best way to protect Venice, and on December 6, 2001, an Italian commission led by Prime Minister Silvio Berlusconi approved the proposal to build the gates, with Italy's Environmental Ministry promising to take the ecology of the lagoon into account at every step. The final obstacle had fallen. "The amount of money needed—$2.3 billion—is not forbidding," says Gentilomo. "We spend crazy amounts of money on football stadiums and many useless projects. Each square meter of flap gate should cost around $50,000, comparable to costs of the Oosterschelde Barrier. London's Thames Barrier cost four times as much." The preservation of one of the most astonishing monuments of the Western world seems worth the $2.3 billion required to help preserve it for coming generations.

After criss-crossing the Venetian Lagoon to look at work in progress, and meeting some of the stubborn Italian engineers who fought a hard battle to build their gates and save the city, I thought again about the aspiring disposition of human nature and went to look at the perfect paintings of Giovanni Bellini; the sublime church of Santa Maria Gloriosa dei Frari; Venice's myriad arches, frescoes, and sculptures; and finally the incomparable Piazza San Marco. Against earthbound notions and common sense, men once dared to build such a city. Its survival now depends on the skills of our engineers and the boldness of our politicians.

I asked Maurizio Gentilomo what would happen if the sea-level rise were to be greater than projected. "We are certain that with gates to protect the city, Venice will be safeguarded for the next century at least," he says. "In the case of a substantially higher rising of the sea, perhaps these gates will not be enough."

Gentilomo looks at me pensively. "Sixty percent of the world's population live in coastal areas," he worries out loud. "Industrial and agricultural activities are concentrated here as well. If the sea rises more than one meter in one century, Venice will be only one small problem of the worldwide community. It will mean a transformation of the earth." Gentilomo is right. If the sea rises higher than a meter in the next century, it will be a disaster of global proportions.

Chapter 2

A SIN OF SCALE

The Great Projects

Time alone can tell whether the boldness of engineers is justified.

WILLIAM WILLCOCKS, 1911

One of my favorite stories about water control involves my husband's resourceful Uncle Marion. Several decades ago, he and Aunt Ruth bought a nice piece of land in Washington State, on which there were about three hundred acres of low-lying scrub brush, a handful of trees, an abundant creek pouring off the mountain, and a small swampy area, which maps sensibly identified as "Lost Lake," since the former lake was certainly gone. Marion, however, thought he'd like that lake back and began hacking scrub. He immediately got into difficulty with state officials for cutting trees, which, they felt, might increase water runoff on the floodplain below, where people persistently built new houses. After wrangling over environmental impact assessments and hiring a consultant, Marion finally got his permit to cut, then decided that he'd had enough bureaucracy. Three fine, sprawling ponds now grace his property. I asked him if he needed a permit to create them. "Well, if you were to put in ponds, you would," he says happily. "But I didn't—I put in beavers!"

Marion brought four beavers to live on Lost Lake. Within a year, they had built three dams out of mud and sticks. Today, the ponds run from 20 to 75 feet long and are flush with ducks, birds, fish, and several generations of beavers. The beavers are such good hydrologists that in a year of low rainfall they keep the main and lower dams intact but breach the middle one.

The beaver-crafted lakes are also good for the people living in the floodplain below, since they prevent runoff, even though nearby McCorkle Creek continues to flood them out about once every ten years. When the county decided something must be done to protect the floodplain homeowners, Marion drove down to the county commissioner's office and suggested that they didn't need to spend $500,000 on building a dam. They merely had to put in beavers. The concrete impoundment the county opted to build instead demonstrates a certain lack of imagination, but the creek is controlled, upstream and down. When given a choice between big solutions and small ones, on the whole, people still usually choose big.

HYDRAULIC ENGINEERING is an ancient art. We think of the societies that have grown up along venerable, ancient rivers as the great, water-driven civilizations. Yet it was the twentieth century that saw the largest expansion of hydraulic control throughout the world—an explosion of cement and concrete, man-made dams, lakes, and rivers, power plants and pumping stations. We've been immersed in the building boom long enough to learn something about how things work; yet the bulldozers push on imperiously. Some of these dams may be needed, but many are not.

Dam building has slowed in America where there are about 5,500 big dams and more than two million smaller ones. There aren't many optimal canyons left here, because most good sites already hold dams. It is now more likely that a dam will be torn down than that a new one will go up. Although dam construction may have slowed generally in the

Western nations, dams are multiplying like mushrooms in South Asia, China, and South America, growing taller and wider in ever-larger clusters. Dozens of dams are in the works along the Mekong Delta in Laos, Vietnam, Thailand, and Cambodia. In Nepal, the world's highest dam, the Pancheshwar, is soon to be thrust into the Mahakali Valley; at 1,033 feet it will be just a little shorter than New York City's Chrysler Building. Pumping projects are under way too, as countries like Libya, China, and Spain divert water hundreds, even thousands of miles through massive canals and pipelines. The world's biggest dam-and-tunnel project, the Lesotho Highlands Water Project, will soon carry large quantities of water to South Africa from behind six Lesotho dams, including the gargantuan Katse Dam.

Colossal engineering works bestow big contracts and big benefits, divide up waters, hold them fast, channel them away from some and give them to others. Dams therefore remain some of the largest investments made by any country. Because water schemes are a major component in the annual budgets of many nations, it has always been politics that start the bulldozers moving. Most of the big projects are in the southern hemisphere, yet funders and builders still seem to come from the north.

The question of whether or not a dam or canal should be built has become both more complex and more serious as the demands for water escalate—from cities, from industry, and from farmers, the most vulnerable members of the world economy and the biggest water users. More than a billion people depend on crops grown under irrigation supplied by reservoirs. In some places, if no reservoirs are built, poor people will be denied the means to improve their lives. There are 1,600 big dams now going up around the world. Understanding the complexities of hydraulic control, how and why dams get built, and who builds them, matters more than it ever has.

THE WORLD'S OLDEST BIG DAM, Sadd el-Kafara—the Dam of the Pagans—lies crumbling in the Egyptian desert as it has for nearly five thousand years. Sadd el-Kafara was madly ambitious, 348 feet long, 276 feet wide at its base and 200 at the crest. Made of 100,000 tons of masonry blocks with a gravel-and-stone center, it was built sometime between 2950 and 2750 B.C., its reservoir meant to catch and hold 20 million cubic feet of rain. We can learn a lot even from a dam 5,000 years old. Of tremendous bulk, Sadd el-Kafara was badly built. No mortar fused its limestone blocks, and if there was a spillway at all, it was too small. Although its center is gone, the fact that no silt was found anywhere behind it proves it was not in use very long. Built to service an alabaster quarry near Helwan, it was never used for irrigation, so its lessons are largely structural. No other large permanent dam with a reservoir was built in Egypt until the twentieth century, and none was needed, given the abundance of the Nile.

The Great Dam at Mar'ib in Yemen is another matter. Its pieces lie in a stretch of golden desert, several hours' drive through high volcanic mountains and lava-strewn ground, southeast from Sanaa. I climbed on its ancient sluice gates and happily ran my hands over worn inscriptions and silky mortar—unique, I was told by a Yemeni friend, because it was made with eggshells. The 2,000-foot-long dam at Mar'ib was first built in 750 B.C. The Mar'ib Dam was of considerable mass—once, when it had been washed away, twenty thousand men toiled to raise it again. Between thirty thousand and fifty thousand people once lived here, sustained by the dam for a thousand years. It was raised from 13 to 23 feet in 500 B.C. As ships replaced camels and the spice and incense trade drifted away to sea routes, commerce was weakened in the Mar'ib. The dam was neglected and finally washed away in 575 A.D.

There is not much left above ground of the dam or of the Sabaean civilization that built it—a handful of dizzyingly tall pillars at the Temple of Bilquis and the ruins of the city of Mar'ib, once queen of the desert caravans. When I first visited the ancient dam site in the early eighties, the desert was empty except for a handful of Bedouins and five families living among Mar'ib's ruins. As with Sumeria and Harappa, when dams and water control failed, so did the crops and the strength of armies. Now that a new dam has been built upstream, people have returned to the area, and once again green land edges golden sand. Yet no one imagines that the new dam will last anything like a thousand years. When the new dam at Mar'ib goes, the people of Mar'ib will once again go with it.

A dam can be as simple as a large piece of rubber or an engineer's dreams writ large in cement. It is sometimes a mountain of rubble and sometimes a confection of gates, curves, and tunnels. It may be stirring in its audacity or a scar on the landscape. As a work in progress, a dam is an event, a spectacle, an augury of destruction and disruption or perhaps salvation. These days its construction is nearly always a contentious matter with advocates and proponents bitterly divided into rigid, combative factions. A dam can have many beneficiaries—bureaucrats, multinational banks and businesses, as well as ordinary citizens in need of less or more water. And its effects are often unintended, occurring long after the men who built it are in their graves.

WILLIAM WILLCOCKS, the British engineer who built the first dam across the Nile, loved moving water. He was thrilled by the urgency of rising and falling rivers. He learned to probe and measure their depths and changeable nature as well as anyone else in his time and better than most. When he was twenty-four, the young British engineer saw the sea for the first time. Although he'd never before laid eyes on such a large body of water, the expanse of the Arabian Sea left him unmoved. "The sea did not appear to me half as impressive as the desert or as a river in flood," he mused. "It only struck me as being unutterably restless"—too restless, and too big to do much with. The passion of William Willcocks's life was to control water, to contain it, divert it, and push it onto dry land where it would make things grow. "It goes along the line of least resistance," he wrote. "Water cannot be left to itself."

Willcocks was one of the world's foremost hydraulic engineers, a member of an elite colonial corps that vastly extended the boundaries of what could be done with earth, concrete, and water. He was born in

1852 in the foothills of the Himalayas in India, in a tent alongside an irrigation canal under the charge of his father, Captain Willcocks of the Bengal Horse Artillery. Captain Willcocks had seen service in the Punjab and afterward, like many soldiers of the British Raj, had taken up civil engineering. By the age of ten, William was already familiar with the problems of silt in canal locks and securing canal floors against the pounding of falling water. His heroes were irrigation officers—"a singularly devoted body of men," he remembered.

Britain had poured considerable effort into public works in India in the nineteenth century—roads, bridges, railroads, as well as waterworks—but since the engineers had little firsthand experience with irrigation at home, they often botched the job, sometimes with horrifying results. The West Jumna Canal, site of William Willcocks's birth, for example, was originally built by the Turkish Sultan of Delhi Fīrūz Shāh Tughluq in the fourteenth century but had been poorly reconstructed by the British so that it cut across natural drainage lines, which meant bogging and mosquitoes. Of the ills visited on India by water control, the most terrible was malaria.[23] "It was the finest station in India," wrote British traveler Emma Halliday while visiting the former British cantonment at Karnal on the West Jumna Canal in 1854. "But a canal was opened there and then during two years the people died by hundreds, so it was left. We entered by the remains of the church. There are two cemeteries quite full of tombs." While irrigation systems didn't create malaria, they did create seepage, raised water tables, and interrupted drainage—all of which meant standing water and mosquitoes.[24]

But the canals were beginning to pay for themselves, and so malarial deaths in perhaps a hundredth of the population of a canal region were weighed against the benefits of irrigation, lives saved during droughts, the sheer size of the canal systems to be managed and, of course, the profits. Sustained by their successes, the British began building the most daring engineering work of the time, a massive canal to capture water from the Ganges River as it gushed out of the Siwalik Hills. Engineers spent eighteen years building the thousand-mile-long Ganges Canal, the largest canal system in the world.[25] The $15 million canal discharged 6,750 cubic feet of Ganges water every second, sending it barreling down artificial drops, under and over torrent beds, out of the Himalayas and 890 miles across the searing plains.

William Willcocks acquired his engineering skills at Thomason Civil Engineering College at the headworks of the Ganges Canal at Roorkee, where a pioneering aqueduct lifted canal waters high above the Solani River on fifteen soaring arches, each one deep and wide enough to accommodate boat traffic. He later said that he had been so inspired by those engineering works that he turned down the offer of a fully paid education at Cambridge. Upon graduation, Willcocks was sent to the lower Ganges Canal. He took to the field with vigor, and soon married the lovely niece of his superintending engineer, a woman amazingly well suited to his peculiar life. "Camp life in winter on one of the upper India canals can not be beaten," Willcocks wrote. "My wife took the keenest interest in the work and through the whole of the hot weather did her share of the inspection just as though she were a canal official. She rode down one distributary recording facts in her notebook while I rode down another or we rode on different banks of the same canal. We met at mid-day at some inspection house. On one occasion she had sunstroke but I was able to get ice by train and pack it round her head. After a day's rest she continued her work. We were encouraged by seeing the splendid appearance of our irrigated fields in the midst of the general desolation."

Willcocks's extensive writings are a good look into the mind of a British engineer in the East. "That which is called the white man's burden is in reality his Christmas box," wrote Willcocks. "At any rate for every burden to be carried there are scores of eager applicants. The real white man's burden . . . is to leave the soft life of the cities and in imitation of the old pioneers, to replenish the earth and subdue it."

The British engineers would show again and again that their ability to

move earth and water extended beyond their understanding of its consequences. Water barreling down calamitously steep slopes scoured channel beds, undermined bridges, and eroded embankments. Where water
slowed, silt loads lost suspension, dropped into bends, and clogged
channels. The gradients and falls of the Ganges Canal were so steep that
it couldn't take a full flow without the velocity of the water devouring
the canal structure. Learning from their mistakes, colonial engineers
designed canal headworks to offset heavy silt loads by balancing silt and
scouring. They designed more gradual slopes and learned to flush and
drain waterlogged and salted land, which greatly moderated the incidence of malaria. They rerigged the Ganges Canal and realigned the
West Jumna so that during the drought of 1876–78 it irrigated 395,000
acres of land, made a profit of eleven percent on its costs, and restored
life to the malarial death trap at Karnal.

British dams and canals profited the Crown, but they were also a
hedge against drought and starvation in a place where famine tables
make sobering reading. Ten million died in the famine of 1770 alone,
two million in 1860–61, four million in 1876–78, and more than a million in 1877–79. According to the commissioner at Agra, even in the
especially severe famine of 1896–97, which took no fewer than five million lives, it was in those places without canals where "the pressure of
famine made itself severely felt."

Nowhere else had such a water resource been in the hands of a colonial power with the means to develop it on such a scale. According to
Khuswant Singh, who has chronicled Sikh history, those who colonized
the once barren wastes of the Punjab Canal Colonies "became the most
prosperous peasantry in Asia. The Punjab had become the 'granary of
northern India' and its capital, Lahore, the 'Paris of the East.'" In the
end the British built or refurbished some seventy-five thousand miles of
main canals and distributaries, watering the most valuable farmland in
India.

"Some of these canals were in themselves feats of engineering by which a people might be content to be judged," wrote the veteran civil servant Phillip Mason in *The Men Who Ruled India*. "It is an idea to stir the dullest— a desert ready to be peopled, a Utopia waiting for its architect. And there is something staggering about its success." The works were not all successes but India was a great proving ground, and the men who learned their craft watering its dusty surface eventually moved west to Africa and the Middle East eager to bring more sun-baked earth into submission. William Willcocks was among them: "I had lived in India 31 years," he remembered later, "and loved the country and the people, but I was pleased to get an opportunity of seeing other parts of the world." His first assignment was a troublesome barrage in Egypt.

"Green, inexpressibly green is the vale known as the land of Egypt," wrote an early traveler. But Egypt's green is a trick and a deception, because this land is almost entirely without rain. Life here is born out of water and silt that has traveled thousands of miles from the rain-fed highlands of Ethiopia and Central Africa. Egyptian civilization was built around the manipulation of these waters. For millennia, Nile waters had come roaring down the riverbeds in late summer, and as the floods rose, farmers channeled them through canals into large basins where they saturated plots of dry farmland and deposited rich silt. But the Nile was unpredictable. When there was too much water, the destruction it caused was terrifying. When the floods failed, the drought and famine that followed were murderous. Since ancient times, Egypt's rulers had been confronted with the difficulties of ensuring water for crops while protecting the country from floods. An ancient hieroglyphic figure that signifies "province" depicts a segment of an irrigation ditch. "The Egyptian question *is* the irrigation question," said a nineteenth-century Egyptian prime minister.

William Willcocks arrived in Egypt in 1883, two years after the battle of Tel-el-Kebir, in which the British, interested in protecting eastern routes through the Suez Canal, somewhat reluctantly squashed an Egyptian uprising against the Ottoman Empire in about two hours, and found themselves with a country to manage. Willcocks was made inspector of irrigation under Consul-General Evelyn Baring, the Earl of Cromer, who, having witnessed firsthand the economic benefits of irrigation investment in India, promptly began raising money to retool Egypt's irrigation systems to send more cotton to the mills of Manchester.[26]

A mature William Willcocks, stubborn, unorthodox, with little tact and no tolerance for toadying, never bent easily to Cromer's authoritarian ways. But he loved his work and quickly became preoccupied by the threat and promise of the Nile. "The Nile looms very large before every Egyptian and with reason," he wrote. "I remember well the feeling of awe I experienced when in August [of 1884] I saw the river coming out of the deserts . . . rising and swelling until it overflowed its valley, while overhead we had a cloudless sky under a burning sun. I at once realized why all Egyptians, like their earliest ancestors, believe that on a certain day every year, an angel from heaven descends with a drop of water from near God's throne and lets it fall into the Nile."

Willcocks began his Egyptian career by repairing two shoddily built French dams in the Nile Delta, which leaked, breached periodically, and were the cause of much misery and countless deaths. Willcocks fortified the dams and, by maintaining a higher water level both behind them and in the canals they fed, reduced the need for labor, thereby dispensing with the corvée, the brutal system of conscription by which Egypt's rulers had maintained the canals for thousands of years.[27]

Willcocks lived in a houseboat on the Nile at Helwan and there developed a lifelong concern for the people along its banks.[28] He came to believe that engineering could solve all problems having to do with water and that while those solutions were vital to the well-being of all people, they were especially important to the poor. It was at Helwan that he began to write about the complications of engineering works that would cause so much distress in the later part of the twentieth century. He foresaw the problems of salt and insisted that deep drainage was the key to irrigation. If ground water rose above a certain height, land would become waterlogged, he argued, and he insisted that canal water levels should be immediately lowered and lift irrigation imposed—the use of manual or simple lifting devices to raise small amounts of water rather than the large amounts discharged by canals. He believed in moving steadily but with care, because, he said, "moving quickly and lifting quickly is not engineering."

The Nile was still an enigma. The British explorers—Burton, Speke, Baker, Livingston, and Stanley—had captivated the world with their search for its sources, but when Willcocks arrived in Egypt the river's depths and flows remained as unknown as its sources had been. Willcocks found the mysteries of its behavior just as enthralling as its geographical wanderings. Nevertheless he was determined to find a way to protect Egypt from both floods and famines. He would dam the Nile. No one had ever dared try to dam a river of such width or ferocity.

Willcocks set out in 1890 to look for a suitable site for a dam that would hold back an immense reservoir. For three years he and his men surveyed, sounded, and made cross sections of eight hundred miles of the Nile from the delta into the Sudan, sleeping each night under the open sky. "It was easy to understand why the Arabs used to say that they knew only one illness, the illness of death," he wrote. "The clear atmosphere, the bright starry night, the sunshine, and the bracing air were enough in themselves to make one thank heaven daily for such blessings, but when to them was added the knowledge that we were working at one of the great projects of our time, I can truly say that I often felt, as I did afterwards in Mesopotamia, that I was not walking on this prosaic earth, but was being borne on wings from place to place."

Willcocks became convinced that there was only one place for such a reservoir: at the caravan town of Aswan. He then began to design a dam unlike any other, a gravity dam, which would resist the weight and lift of water by virtue of its own bulk, but curved to fit the half moon of solid rock underlying Aswan's riverbed. At 6,400 feet, it would be the longest dam ever built, its masonry facade penetrated by 180 sluice gates—140 along the bottom of the dam and 40 on the top—through which the Nile in full flood could flow. After the heaviest waters had passed, the gates would close to catch and store water. The sluice gates were crucial since Willcocks understood if the great silt load that washed down from the Ethiopian highlands wasn't allowed to move through the dam, it would build up, raising river beds, diminishing the capacity of the reservoir, and depriving the delta farmlands of the river's gifts.

Willcocks's daring design was approved. But by then—the mid-1890s—he had begun to run afoul of Lord Cromer, and lesser men were promoted over his head.[29] Then war in the Sudan stalled the building of the dam, and in the spring of 1897, Willcocks, believing that his dam at Aswan would now never be built, left government service in disgust. But he had underestimated the British drive for profits. If they were to make money from cotton, a summer crop that demanded water when the river was at its lowest, they needed a year-round water supply. Work on the dam was begun in December of 1898 by the British firm John Aird & Company, using Willcocks's design—but not its designer.

The first and most daunting challenge was to close the channels of a wily river that all too often sent cataclysms roaring down its bed. Temporary stone dams were jammed across three of the Nile's five channels so that sandbag blockages could be laid. Stone dams of boulders and *shimfs*—wire cages filled with small stones, each set weighing from one to four tons—were put in place by cranes. Smaller stones were tipped from wagons into the rippling currents.

At first, Nile flooding was blessedly low, so the job seemed easier than expected. It took just three months for workmen to fill, then close, the first two channels. But because the river's flow had been narrowed and intensified by the closure of the first channels, the third channel had become treacherous. Stones weighing three and four tons were swept downstream like pebbles. Finally, workers loaded two railway wagons with wire and stone *shimfs,* each weighing twenty-five tons, and dumped them into the rushing waters, forming a "toe" against which more rock was piled. The last channel was closed in July 1899.[30]

An international commission of advisors, alarmed by Willcocks's unconventional design, had taken out its lovely curve. This proved to be a costly mistake because, without that curve, men excavating the foundation hit granite so soft it could be chiseled with a pick. Willcocks had carefully curved his dam to coincide with the solid bedrock on which it needed to sit. To compensate, engineers had to dig out the soft rock, grout cracks with cement, and lay a concrete foundation on top. A masonry foundation was put in place. On just one record-breaking day, 3,600 tons of masonry were put down.

It was a project worthy of the pharaohs. The interior of the dam was made of clay and limestone mortar. Italian stonemasons—eight hundred of them—were invited to Egypt to cut locally mined granite blocks into pieces that would fit into place on the dam's exterior. The blocks were then transported by train to the banks of the Nile, loaded into boats, sailed to the dam, and then hauled into place by hundreds of men. One and a half million cubic feet of masonry were placed by hand.[31]

Willcocks was still in Egypt, having taken a job as manager of the Cairo Water Company. It must have been almost unbearable for the engineer to watch his greatest work go up without having a hand in its construction, and agonizing to watch its builders deviate from his plans and plow the dam directly into soft rock. Their decision to use a plain sill rather than the ornamental pylon cornice, which he had designed to be as grand as the surrounding temples, especially outraged him. "The dam with this ridiculous cornice can only be compared to a king clothed in ermine and gold," he wrote, "seated on his throne, with a bundle of groceries on his head."

The greatest controversy generated by the dam at Aswan arose over the ancient temple complex of Philae, which would be submerged for four or five months out of the year by the reservoir. The young journalist Winston Churchill weighed in on the matter. To halt the dam's construction, he said, would be "the most cruel, the most wicked and the most senseless sacrifice ever offered on the altar of a false religion. The state must struggle and the people starve in order that professors may exult and tourists find some place on which to scratch their names." Finally, it was decided to keep the dam at a height of sixty-five feet, which would make the reservoir lower so that the complex would be only half-submerged. Thousands of Nubians upstream of the dam were displaced with less concern.

Willcocks's stately, handsome dam at Aswan was completed in 1902. Never before had such a mighty river been caught and held. The reservoir behind it stretched two hundred miles into the Sudan and held a billion tons of water, enough to green the fields of Egypt even in a poor year. "A spark has fallen into the primeval world," wrote the German traveler Emile Ludwig after seeing the dam. "This is the great construction by which men mastered the element, the bold invention which has determined the fate of the Nile upstream and downstream. It is the point at which its freedom ends."

Whether or not the dam was a spark in the primeval world, it was, as Ludwig also called it, "an epoch marking event," one of the finest dam-building achievements of all time. Egypt was no longer subject to the uncertainty of seasonal flow but could draw upon stored water all year round. The amount of Egyptian land under irrigation increased by half a million acres. The economy boomed. But soon it would not be enough. When the population grew and more water was needed, Willcocks's visionary design allowed his dam to be raised twice, first in 1911, then again in 1934, adding a total of fifty feet and enough storage to irrigate another million acres.

William Willcocks had changed the way men thought about water. For his service to the British Empire, he was knighted. During his tenure with the British government, Willcocks surveyed for the first time all of Egypt's riverine landscapes and traveled up the White Nile to the Equatorial Lakes, walking the final four hundred miles to Entebbe. He knew that the real key to control of the Nile lay upstream, and dreamed glorious dreams of other engineering works: a massive storage reservoir at the source of the Blue Nile in Abyssinia, dams in the Sudan and at Victoria and Albert Nyanza lakes, and the restoration of a reservoir at Lake Moeris in Egypt that he believed had been created by twelfth-dynasty pharaohs. In 1913, he published *Egyptian Irrigation,* a two-volume work detailing everything he and other engineers had learned about the Nile, filled with delicate drawings of geographical formations, weather systems, locks, bridges, and barrages. At the invitation of the Grand Vizier of Turkey, Willcocks drew up a plan to reclaim the land between the Tigris and Euphrates rivers, which included myriad dams, the renovation of Persian canals at Nahrwan in Iraq, and the diversion of flood overflows into large natural depressions in the desert. His plans were the basis for work that went on for decades after his death.

When he died in 1932, an obituary in the *Diocesan Magazine* of Egypt paid tribute to his lifelong humanitarian work for the people along the Nile: "And that pasha, who knew the Egyptians, who helped to abolish their corvée and give them their Nile, and improve their health, their wisdom, their diet, their agriculture, their common sense, this pasha . . . in 1932, departed this life to the great glory of God."

The *London Spectator* paid homage to the things for which he would be remembered in the world at large: "Sir William Willcocks, whose death at the age of 80 we record with regret, built his own monuments in Egypt and Iraq, as durable as those of Cheops and very much more useful."

THE HISTORY OF dams and canals has been an ongoing story of works that grew bigger, longer, and taller as each achievement paved the way for a greater one. The first large-scale experimentation with water control took place in the eastern colonies of the British Empire, but it was the opening of the American West that would revolutionize hydraulic engineering. Around the turn of the twentieth century, half the continent waited to be exploited—open, seemingly endless expanses of land. Still most of that land was parched and subject to harsh weather. Settlers and explorers isolated on prairies, in mountain valleys, and rainless deserts, figured out quickly how to store and move water, techniques born out of frontier necessity. What they learned moved the field of engineering forward dramatically.[32]

As William Willcocks was designing the dam at Aswan, there were already forty thousand engineers in the United States, and they were developing a distinctly American character, particularly on the frontier. They were brash, adaptable, and imaginative. After a 1914 visit to dams and irrigation works in the American West, Willcocks himself remarked on the distinctive spirit of his colleagues: "The modesty with which all the engineers spoke of their work, some of them really colossal undertakings, has made a marked impression on me . . . All were so impressed with the greatness and potentialities of their country that their own personal efforts seemed like pygmies."

Congress had passed the Reclamation Act of 1902 to advance western settlement by storing and moving water, and in the early days of dam building, records began falling fast. When the 328-foot-high Buffalo Bill concrete arch dam was completed in 1910, it was the tallest in the world. Five years later, the Arrowrock on the Boise River in Idaho dwarfed it at 348 feet. These dams began to transform the western landscape, many bearing the wild and romantic names of their surroundings: Minidoka, Umatilla, Lahontan, Pacoima, Hetch-Hetchy, Pathfinder, Belle Fourche, Elephant Butte, and Wind River. The West was dotted with dams, some lemon-slice thin and strangely shaped, flaunting domes, cupolas, vertical slabs, and steps. Americans pioneered several kinds of structural dams, including multiple-arch and flat-slab buttress dams and lightweight, narrow buttress dams, bolstered by discrete arches—sometimes called "hollow dams" because of the empty space between buttresses.[33] But the dam that would change everything would be built at Boulder on the Colorado River.

The Colorado was so wild that its basin had been the last place in the country to be fully explored by white men. Even so, Americans were determined to bring the Colorado to heel—to stop treacherous floods from pounding down its reaches and to keep it from running dry as well. In 1922, after a lengthy debate over how best to tame the Colorado, Arthur Powell Davis, Director of the Bureau of Reclamation and mastermind of the Buffalo Bill and Arrowrock dams, confidently handed the government a comprehensive proposal to quiet the raging river and make it useful. At its heart was to be a soaring dam, higher by three hundred feet than any on earth.

To build it, a group of heavy-construction companies—Henry J. Kaiser, Bechtel, MacDonald and Kahn, Morrison-Knudson, Utah Construction, J. F. Shea, and the Pacific Bridge Company—formed a coalition that, like the project they aspired to, was unprecedented in size and scope. They called it "Six Companies." These were tough and experienced men, and if they couldn't build the outsize Boulder Dam, nobody could. But even they couldn't have imagined the job without the new machines that were then coming into use—lumbering, powerful, big machines.

The 1910 Buffalo Bill Dam had been built with mules, shovels, and dynamite, but during the 1920s, one writer suggested that gasoline had replaced hay as the empowering fuel of the West. Machinery began to make possible projects on an undreamt-of scale. When the metal treads that kept World War I tanks moving through the mud and soft earth were combined with gasoline- or diesel-powered engines, and scrapers or steel blades, to create earthmovers and excavators, it took only a handful of men to rearrange previously unthinkable quantities of clay and rock.[34]

Concurrent with advances in heavy equipment, engineers and scientists were also investigating hydraulics, geology, soil, drainage, and the expansion, contraction, and thermal properties of concrete itself. A technique perfected on the Owyhee Dam in Oregon—the use of artificial cooling by circulating river water through pipes embedded in the dam's concrete structure—was also used at Boulder. Workmen deployed a wild-looking innovative rig called a "jumbo," a four-level mobile drill carriage that could mount up to thirty rock drills so that thirty men could simultaneously bore into the hard rock of the canyon at record speeds—and noise levels. There wasn't an area of design and construction that the big dam didn't move forward several notches: seismic analysis and testing; the use of special thermometers, gauges, and meters; shrinkage control; model studies of spillways, outlets, valves, and control gates; cement and concrete composition; and stress analysis.

In 1935, the dam, now officially named for ex-president Herbert Hoover, hit its final height of 726 feet, as tall as a sixty-story building. Its penstocks were big enough for a railroad train to pass through.[35] Hoover, which took five years to build, was the first dam to be entirely designed using the trial-and-load method, a devilishly complex form of analysis that would be the basis for building hundreds of subsequent dams. Prior to the building of Hoover, the number of dams anywhere in the world that contained a million cubic yards of concrete could be counted on one hand. After Hoover, dams of such size became almost routine.

"This morning I came, I saw and I was conquered," said President Franklin Delano Roosevelt at the dam's dedication. "All these dimensions are superlative. They represent and embody the accumulated engineering knowledge and experience of centuries; and when we behold them it is fitting that we pay tribute to the genius of their designers. We recognize also the energy, resourcefulness and zeal of the builders, who, under the greatest physical obstacles, have pushed this work forward."[36]

Built during the Great Depression, Hoover Dam was a monument to work, hope, pride, and, most of all, possibility. "Labor makes wealth," said the president. "The use of materials makes wealth. To employ workers and materials when private employment has failed is to translate into great national possessions the energy that would otherwise be wasted. Boulder Dam is a splendid symbol of that principle. The mighty waters of the Colorado were running unused to the sea. Today we translate them into a great national possession." The great monolith in the desert canyon was not just a grand engineering accomplishment, it was a political triumph. Its sheer scale seized the imagination, not just of the continent but of the world. Was there anything that we couldn't do with our machines, our science, our daring? The idea of reclaiming wasted waters in the service of man would control popular thinking about water use for decades to come.[37]

Encouraged by senators and congressmen and the people of the United States, the Bureau of Reclamation built the world's largest dams and irrigation projects for half a century all over the West—355 dams with reservoirs, 254 diversion dams, 16,000 miles of canals, and 52 hydro plants.

Hoover Dam transformed the Kaisers and Bechtels and Morrison-Knudsons into heads of international corporations, making rough-and-ready dirt movers into tycoons with an appetite for big jobs. Even before it was finished, they were looking around for other work. The jobs were there, thanks to Franklin Roosevelt and his New Deal. Big dams provided flood control, electricity, irrigation, and employment. Roosevelt's daily notes to Secretary of the Interior Harold Ickes read like shopping lists for projects and numbers: The Santee-Cooper Development, Skagit, Truckee River, Carson River, Humboldt, Ruby Dam. "How about it?" he queried Ickes in his sprawling handwriting. "Can we survey? Can we take money from here and put it there? Can we do it?" In 1936, the four biggest dams ever built—Hoover, Bonneville, Grand Coulee, and Shasta—were all under construction in America, and Six Companies was involved in them all. "No group of Americans in history have spread themselves out on the scale of these Westerners," wrote *Fortune*.[38]

Soon the influence of American builders began to be felt around the world. They were pleased to hear themselves called "ambassadors with power shovels." Utah Construction built ports in Peru and Mexico and dams in Colombia, France, Australia, and Pakistan.[39] Morrison-Knudson worked on the Saint Lawrence Seaway, built dams in Canada, Labrador, Iran, Mexico, Turkey, Afghanistan, and Australia. Kaiser worked in more than fifty countries, from Africa to Australia, from Alaska to Brazil. "Nothing is too big or costly to dream about or to study seriously," said a Bechtel engineer. "Somehow, sometime—if the need exists, the result will be accomplished even though today it appears possible only as a joint venture of King Midas and Paul Bunyan." No one seemed to find this barreling expansion the least bit scary.

WILLIAM WILLCOCKS'S DAM at Aswan had been the first great dam of the twentieth century and launched the large-scale manipulation of rivers. But by the time Willcocks's dam was raised for the second time in 1934, the needs of Egypt's rapidly growing population had already begun to outstrip its capacities.[40] Eventually the Aswan Dam would be supplanted by a still more ambitious structure—notorious before its construction and after, and one with immense consequences for Egypt.

The second dam at Aswan was born of the same urgencies that prompted the first: fear of drought, famine, and floods. A dam on the Nile that answered those fears also meant political security, first for the British seeking to solidify their hold on Egypt and then for a nationalist government seeking to establish itself.[41] After the corrupt King Farouk was overthrown by Egyptian armies in 1952, the country's cash reserves were depleted and its population had doubled since the building of the first dam. Independent Egypt's government, led by Gamal Abdel Nasser, was desperate to find ways to produce more food, and envisaged a new dam and reservoir on the Nile as a symbol of the new Egypt—a highly visible, technologically sublime monument in a land overstuffed with glorious monuments.[42]

After Western governments backed away from Egypt, the Soviets—for whom the structure was an emblem of their presence in Africa—agreed to finance the High Dam at Aswan. The High Dam was completed in 1971, at a cost of nearly $1.5 billion. At 364 feet—not at all high—it is an unremarkable, even lumpish-looking, rock-fill structure, but it backs up one of the largest man-made lakes in the world. The whole lower Nile was now under control, its waters captured in a three-hundred-mile-long reservoir named after President Nasser, who died a year before his monument was completed. Although its size has since been surpassed, the High Dam remains a source of controversy. After thirty years, there are still no simple answers to whether or not the High Dam is a savior or monster.

Lake Nasser loses as much as fifteen percent of the Nile flow to evaporation in a year, and more to seepage around its edges. Downstream fields have been afflicted with increased salt and waterlogging. The loss of the Nile's silt loads, so carefully carried through William Willcocks's sluice gates, has meant the accumulation of silt in the reservoir, two hundred kilometers behind the dam. Meanwhile, what Willcocks called a silt-famine has hit downstream, depriving both brick makers along the shoreline and farmers who have been forced to replace the silt with dangerously large amounts of chemical fertilizer. Because the silt—110 million tons of sediment a year—that built the Nile Delta over seven thousand years is no longer replenished, the river's banks are eroding and saltwater pushes farther and farther back into the Nile Delta. Sardines no longer come to feed in its silty waters, so the fishery has been wiped out.[43] Because the Nile is no longer flushed by yearly floods, water quality has deteriorated under loads of fertilizers, sewage, and industrial waste. Both Lake Nasser and the canals have been deprived of oxygen by outbreaks of plankton, phytoplankton, and foul, suffocating blue-green algae. There was a net loss in the amount of irrigated land but at the same time an increase in that land's productivity. One of the harshest aspects of the reservoir is the spread of disease over the increased areas of standing water, primarily malaria and schistosomiasis, a nasty, debilitating illness caused by the presence of a parasitic worm in veins around the bladder and intestines.[44] Schistosomiasis is treatable, however, and therefore preventable through education.

Resettlement has had mixed success in both Egypt and the Sudan. In Egypt, the fifty or sixty thousand people who agreed to leave to make room for the reservoir were mostly relocated to newly irrigated land at Kom Ombo, where despite leading healthier lives they have been unhappy about living in the desert. Nubian peoples in the Sudan—53,000 of them—were again dislocated. Moved to a newly dammed area on the Blue Nile at Atbara, they found their physical conditions changed for the better but complained that they were miserably far from familiar landscapes and families.[45]

Dozens of nations participated in excavations, surveys, and the moving of monuments in an international rescue effort led by UNESCO, most famously the grand temples at Abu Simbel and Kalabasha, now well-visited tourist sites on high ground. The temples of Philae, a source of controversy in Willcocks's day, were moved to Agilkia Island with $6 million of American money. Many other sites are under water.

Egypt has struggled to contain its problems, adding tile, pipe, and vertical drainage to millions of acres of irrigated fields. In the sixties, engineers introduced a system that recycles water taken from the Nile three or four times. At least 7.5 billion cubic meters of water are recycled yearly in Egypt. Some of the dam's adverse impacts may have been avoided if the dam had not been such an overt declaration of Egypt's independence.

In Egypt itself, the High Dam remains a source of stubborn pride. "Once the decision had been made to go ahead with the dam—in the teeth of Egypt's enemies, as it were—there was no going back," writes Nile specialist John Waterbury. "The real tragedy lay in the subsequent difficulty in presenting anything but the most positive and most exaggerated benefits to be derived from it. Anything less than superlatives became potentially treasonous."

Outside of Egypt, there has been a decided disinclination to admit that the dam was anything but a colossal error. Yet navigation along the Nile has improved and freight traffic increased. Although downstream fisheries have suffered, fish yields in the reservoir are unexpectedly good. The dam immediately bestowed 4.4 billion kilowatt hours of electricity a year on Egypt, half of the country's output at that time. It rescued Egypt from floods in the great rains of the early 1960s and again in 1975. It saved the country from the drought of 1972–73. During a terrible, prolonged drought that stretched through most of the 1980s, the High Dam continued to provide water for farming until Lake Nasser was almost dry, much of the Nile bottom exposed, and riverboats stranded. Then, in 1988, just as water releases were about to be halted altogether, heavy rains arrived and were captured by the reservoir. Drought returned the following year and the High Dam once again saved the country.

How do you assess the importance of water security in a desert country with just one source of water? The haunting photographs made during the 1984 famine in Ethiopia that killed half a million people offer a stark hint of what Egypt has been spared. "The real question is not whether the Egyptians should have built the Aswan Dam or not—for Egypt realistically had no choice," says Dr. Mostafa Kamal Tolba, executive director of United Nations Environmental Program and World Congress of International Water Resources. For Dr. Tolba, the real question is what steps should have been taken to reduce the adverse environmental impacts to a minimum.

Africa expert Robert O. Collins agrees but for another reason: "Politically, it was a gigantic and daring scheme, a monument to the vision of the revolutionaries," he says. "Economically, it provided water and power. Most important, and the decisive argument before which all the sophisticated, mathematical proposals of British hydrologists for the rational development of the Nile Basin crumbled to dust, the High Dam would free Egypt from being the historic hostage to upstream riparian states by providing over-year storage within the boundaries of Egypt. Whatever the demerits of the High Dam, they were rendered insignificant by that fact."

AMERICAN ENGINEERS working in the West fondly remembered William Willcocks's 1914 tribute to their boldness and audacity, but few recalled the warning that had gone with it: that daring new dams then being built on the rivers of North America were bolder than those in countries of the East because the West had not experienced the tragedies that had occurred in arid eastern lands where so many had perished as the result of man's interference with nature. Once the population of the West had grown beyond a certain point, he said, American engineers, too, would show more caution. "Where the welfare of millions of people is concerned," said William Willcocks, "we cannot afford to run any avoidable risks."

When Willcocks's dam at Aswan went up, a hectare of land—two and a half acres—supported two people. Just before the High Dam was built, that same hectare had to support seven people. At the beginning of this century, a hectare of land must support twenty-eight people. "The population of Egypt is increasing rapidly," wrote William Willcocks in 1913. "The rate of increase is greatest where irrigation projects have been most active."

In America, it took the Bureau of Reclamation a few decades and a monumental amount of cement and stone to make possible the large-scale settlement of the American West. Almost no one lived in Nevada when the Hoover Dam was built, nor were there many people in New Mexico, Wyoming, or Arizona. Hoover changed everything. Hoover's Lake Mead—which could store two years' worth of the Colorado's flow—endowed the entire Southwest with electricity, flood control, and water. More than half of the people who now live in the West depend on water from the Colorado River, which, before it runs its course, has been drained dry.[46]

Hoover itself, although it spawned a generation of extravagant water-works, is not a disaster. It has been described by Larry Stephens, the executive director of the United States Committee on Large Dams, as a monument whose benefits outweigh its costs. Even some anti-dam people believe it to be a "good dam," "not a problem." Unless, that is, you consider Las Vegas a problem.

Las Vegas was a tiny, bawdy gambling settlement before the Hoover Dam was built. But almost as soon as construction began, it became an entertainment center for twenty thousand dam builders and the thousands of tourists who poured into the area to have a look at the dam. Without Hoover to supply electricity to run its air conditioners and water to fill its pools, Las Vegas would not have become anything much more than another little gambling town. Hoover was the catalyst in the transformation of a small strip of casinos into a water-sucking desert nightmare.

Most desert cities have a brave, dusty appeal, at least in part because they so clearly defy nature, but in no other bone-dry city in the world do the lights sparkle as persistently and provocatively as in the scorched air of Las Vegas. Driving along the Vegas Strip one can witness some of the most extravagant water madness anywhere. Not far from the pyramid of the Hotel Luxor are the five-story waterfalls of the Mirage. With its shark tanks, 1.3-million-gallon dolphin pool, and make-believe volcano, the Mirage uses almost a million gallons of water a day. Next door, Treasure Island is circled by a man-made river into which a full-sized pirate ship is sunk by a British frigate, again and again. The Bellagio sits astride an eight-acre faux lake with hundreds of fountains spitting two hundred feet into the air. All around the cartoon city are water parks, dancing fountains, golf courses, housing developments that boast man-made lakes, and one of the most absurd of all desert conceits—grass. One millionaires' development, called "The Lakes," even has a sailing club. No other desperately dry place flaunts water to the degree that this fantasy city does, its illusion maintained and nurtured by technical wizardry. Every year more than 35 million visitors pass through Las Vegas, leaving behind profits of $6 billion in the hotels and casinos. Las Vegas Water Commissioner Patricia Mulroy estimates that each acre-foot of "decorative water" on the Vegas Strip generates revenues to Nevada of $30 million.

Only 3.8 inches of rain falls on Las Vegas in an average year. That's comparable to Muscat in Oman, Riyadh in Saudi Arabia, Kashgar, Port Sudan, or Villa Cisneros in the Western Sahara. It takes more water by a factor of two or three times to live in the Southwest than anywhere else in our country, mostly because of grass, air conditioning, malls, swimming pools, and golf courses. A householder in Clark County—which includes Las Vegas—uses three times as much water as an average consumer in an eastern state.

Las Vegas is the fastest growing metropolitan area in the United States. To accommodate growth, Las Vegas, which now gets eighty percent of its water through an aqueduct from Lake Mead, needs new water sources and has not been shy about grabbing them wherever they are to be found.[47] There aren't many places to look.[48] Las Vegas has applied for more from the Colorado, hoping to see a renegotiation of the Colorado River Compact, which apportions each state's share. It has also applied for water from Zion National Park's Virgin River in northern Nevada and bid for unused water in twenty-eight "waterlogged basins" to its north and east to be transported via a twelve-hundred-mile pipeline.[49] The water in those basins has taken centuries or even millennia to accumulate, and its removal is called "mining," since in such a rainless place it will not be replaced.

"Water," it is often said in the West, "flows uphill, toward money." It may be hard for Las Vegas to resist keeping its money-machine going, but other Nevadans fear that if the city is successful, the state's rural landscape will become a dust bowl.[50] Las Vegas's lavish water use has already dried up area springs and wetlands, taking with them several species of poolfish, stands of mesquite, and resident birds. Land has dropped as much as six or eight feet in some places because of the overdraught on aquifers, forcing residents to abandon some buildings.

Las Vegas epitomizes an unchecked expansion, which threatens to roll over everybody in its way. The spectacle of overuse may be most dramatic here, given the city's gaudy nature, but the West is peppered with growing water crises. California alone will add another 13 million people to its population by the year 2020. The Southwest is booming, hundreds of thousands of people drawn to Sunbelt cities by the region's warmth, dry climate, and beautiful desert. In Phoenix, already home to 2.5 million people, conservationists say that new projects gobble up desert land at a rate of an acre an hour.[51]

In 1878, John Wesley Powell, the first eastern American to explore the Colorado River Basin and the Grand Canyon, tried to make it clear to Congress in his "Report on the Lands of the Arid Region of the United States" that there was only enough water to make about a fifth of western lands habitable by people in large numbers. Powell, who believed in western development but thought it ought to be managed carefully, was discredited but continued to sound a warning. "Gentlemen, it may be unpleasant for me to give you these facts," he told a National Irrigation Congress in 1893. "I tell you gentlemen, you are piling up a heritage of conflict and litigation of water rights, for there is not sufficient water to supply the land." Congress didn't listen. Its members were looking for someone who would tell them what they wanted to hear: that the West should be wide open to development. Powell saw the modern struggle over the West's precious water coming but even he might have been surprised by Las Vegas.

You can take a barge ride on a little Nile River that winds its way beneath the lobby of the Luxor Hotel in Las Vegas, while a grizzled guide will explain the mysteries of the ancient Egyptians to you and the rest of the T-shirted tourists. Outdoors after dark, the hologram of a pharaoh dances above a fountain. Egypt in the Great American Desert! There is an insane logic to it. Egypt and Nevada are among the driest places on their respective continents, and both are engaged in an intense denial of that fact.

Engineers in Egypt began with a careful, well-engineered dam across the Nile and progressed to a larger dam with larger problems. Since I began writing this book, Egypt's population has grown from 52 to 69 million. Egypt, 97 percent desert, already uses virtually all of the Nile's flow to keep its people alive. The High Dam bought it time, but even the Egyptian Nile will soon be strained beyond its capacity.

America started with an array of small dams and quickly moved to monumental structures. Water control in the United States contributed directly to reckless expansion and the overconfident notion that our engineers can eternally dig deeper or scout farther to find rich water reserves. "Bureaus loaded with know-how," wrote Wallace Stegner, "are short on know-*whether*." Our technological entrance into dry lands tricked us into thinking that we could live anywhere we wanted and still have everything we need to make us happy and comfortable. We've known abstractly for a while now that we can't have it all. But we still haven't reconciled ourselves to going without anything.

"If the unrestrained engineering of western water was original sin, as I believe," wrote Stegner, "it was essentially a sin of scale. Anyone who wants to live in the West has to manage water to some degree." Stegner was talking about America, but he might have been speaking of the entire world. The immediate effects of colossal engineering projects are not always ruinous. But one of their most profound legacies—attracting large populations onto land that cannot sustain them—may be more injurious than we thought. The consequences of dams and canals will be felt for centuries after they are created. We ought to be very, very careful what we wish for.

Chapter 3

A Thousand Valleys

River Basins and Utopian Dreams

*We have a choice. There is the important fact. Men are not
powerless; they have it in their hands to use the machine to
augment the dignity of human existence.*

DAVID LILIENTHAL, *DEMOCRACY ON THE MARCH*, 1953

The North Indian state of the Punjab is a bustling place. The roads are alive with cars and buses, horse and bullock carts, bicycles and motor scooters, three-wheeled taxis and herds of sheep, each moving along at their own, highly individual pace. There are tractors here too, carrying handsome Sikh families—black-bearded men in turbans of sizzling colors, their large, beautiful women and irresistible dark-eyed children. Ancient trucks travel the roads, so heavily laden with grain that their burlap sides threaten to touch the ground and sometimes do, taking the trucks over and down. Tall piles of keanu oranges and stacked jars of honey are for sale alongside roads lined with fields of golden mustard, green wheat, and thickets of sugar cane ready for the blade. Red-brown canals curve through the land carrying blue water. The Punjab glows with prosperity.

My traveling companion, Diwann Manna, a local artist, told me that Punjabis have an especially deep sense of belonging to the land and that rivalries over land and over water are passed on from generation to generation. We had driven from Chandigarh, the Punjab's serene capital city, to the gray-blue Siwalik Hills, preamble to the white giants of the Himalayas, to look at the Bhakra Dam, India's first "home-built" post-Independence big dam, and Prime Minister Jawaharlal Nehru's dream.

Overlooking the dam stands an oversized bronze statue of Nehru, eyes gazing down at the symbol of all that he wanted for India—development, agriculture, industry, full stomachs, and entry into the modern world. I confounded my eager guides by bursting into tears at the thought of the authentic hopes and dreams of a new nation that were poured into the walls of this man-made monument to India's prosperous future.

In this time of polarized sentiment about the building of big dams, it has often been forgotten that many dams were built with the best of intentions, by men who genuinely wanted to use them for the betterment of mankind. India's Punjab, the country's most economically hearty state, hasn't always been this prosperous. The Bhakra-Beas System, a group of dams that work together to bring the waters of Himalayan rivers to work on Punjabi farmlands, was born out of the idealism of men like Prime Minister Nehru and the American technocrat David Lilienthal. Lilienthal understood long before most that land and water are inextricably linked and devoutly believed that by developing those resources while conserving them you could change the lives of millions of people. Lilienthal, who brought his message halfway around the world to India and Pakistan, had learned those lessons in the Tennessee Valley.

"Through the long years," wrote David Lilienthal, "there has been a continuing disregard of nature's truth: that in any valley of the world what happens on the river is largely determined by what happens on the land—by the kind of crops that farmers plant and harvest, by the type of machines they use, by the number of trees they cut down. The full benefits of stream and of soil cannot be realized by the people if the water and the land are not developed in harmony. . . . Thus far it is only in the

Valley of the Tennessee that Congress has directed that these resources be dealt with as a whole, not separately."

WHEN THE GREAT DEPRESSION began in 1929, the Tennessee Valley suffered more than most sections of the country. Malaria and malnutrition were abundant but not much else was. The 2.5 million people of the valley earned less than half of what the rest of the country earned and their birthrate was one third higher. Just 300,000 people received electricity, all of it from private power companies. Erosion from wind, heavy rains, and the felling of trees, coupled with poor farming practices, had ravaged the land, and layers of topsoil had long since washed downriver. The Tennessee River itself, flowing out of the mountains of Virginia and North Carolina, across seven states, was wildly unpredictable, with a penchant for flooding Tennessee Valley communities year after year.

Muscle Shoals was a World War I leftover, a little Alabama town on the Tennessee River where the U.S. Army Corps of Engineers had begun to build the Wilson hydroelectric-power dam, a steam plant, and two nitrate-manufacturing plants for munitions. After the war, its disposal posed a problem. George Norris, a Republican senator from Nebraska, argued that Muscle Shoals ought to remain in public hands and be used for the good of the country. Norris was convinced that a revolution could be accomplished in the hard-pressed states of the Tennessee Valley if Wilson Dam were used as a starting point for government-produced power and flood control, and the river opened to navigation. Five times he proposed federal development of the Tennessee River and five times his bills were defeated. When vetoing a 1929 Norris bill, President Hoover said: "The real development of the resources and the industries of the Tennessee Valley can only be accomplished by the people of the valley themselves"—to which one writer retaliated that while this was flattering to the pride of the South, "There is a Southern saying that you can't do nothin' when you ain't got nothin' to do nothin' with."

Then Franklin Delano Roosevelt was elected president of the United States. Roosevelt shared Norris's vision and expanded on it. He wanted far more than electricity and fertilizer from the river. "Instead of doing patchwork here and patchwork there," he told labor lawyer David Lilienthal, "[we could] have a place where you could say to the country, 'This is how it should be done.'"

President Roosevelt signed the Tennessee Valley Act in May of 1933. Its aim was nothing less than the development of an entire region through river-basin control. It would become the world's most ambitious engineering project. The Tennessee Valley Authority (TVA) was created to control flooding along all 650 miles of the Tennessee River, from its network of tributaries in the Appalachian Mountains to its outlet into the Ohio River at Cairo, Illinois. Its goals were to produce and distribute electric power; to make the river navigable by building dams with locks all along its length; to plant trees; to produce fertilizer; to attempt to eradicate malaria; to undertake social and educational programs; to improve farming methods; and to stop migration from the area. The Tennessee Valley Authority was meant to combat poverty itself.

"NEW DAY DAWNS FOR DIXIE," read one Southern newspaper headline just after the signing. "SHOALS BECOMES GENESIS OF 'NEW AMERICA'!" Valley factories blew their whistles in exuberance, expressing the deep hopes of the South. Senator Norris was jubilant: "It is emblematic of the dawning of that day when every rippling stream that flows down the mountainside and winds its way through the meadow to the sea shall be harnessed and made to work for the welfare and comfort of man."

Others weren't so enthusiastic. Critics denounced the Tennessee Valley Authority as "a fantastic pipe dream" and "super pork barrel." Some were genuinely concerned about the erosion of the Constitution; others had their own interests in mind. Labor leader John L. Lewis, for example, worried that Tennessee Valley Authority electricity would deprive his coal miners of jobs. "The gentlemen who are giving birth to this beautiful child," intoned Congressman Everett Dirksen of Illinois, "I am afraid, will not be as proud of it after it grows up." When one Northern writer questioned the advisability of building such a large enterprise in an area relatively free of floods, the Southern writer Willson Whitman countered that he ought to try telling that to the residents of Chattanooga, who had seen steamboats in the streets. "Maybe you never saw a rabbit sitting in a tree," he sniffed. "Maybe you never went to work in a rowboat, and found you had to tie it at the second-floor window of an office building."

To run the Authority, Roosevelt appointed a three-man board of directors, which included a young, bright-eyed, and prematurely balding Harvard lawyer David E. Lilienthal, whose ideas would ultimately shape the Tennessee Valley Authority. Lilienthal's vision for the Tennessee Valley Authority had science and technology at its heart. "There is almost nothing, however fantastic, that a team of engineers, scientists, and administrators cannot do," he wrote. "Today it is builders and technicians that we turn to. . . . When these men have imagination and faith, they can move mountains; create new jobs, relieve human drudgery, give new life and fruitfulness to worn out lands, put yokes upon the streams and transmute the minerals of the earth and the plants of the field into machines of wizardry to spin out a way of life new to this world."

Just three months after Roosevelt signed the Tennessee Valley Authority into existence, the Civilian Conservation Corps began planting trees and building tens of thousands of "gully dams" to cut erosion. Work began on a big dam on the Clinch River in northeastern Tennessee, to be named after George Norris. In 1935, David Lilienthal drove Norris out to look at the big dam. "Norris Dam at night is a spectacular sight under any circumstances," wrote Lilienthal in his diary that night. "But what it must have meant to the old man who had fought for so many years, apparently hopelessly, to establish the principle of public development of the Tennessee River . . . is hard to tell. He said nothing, but his face was a real study."

Within six years, Norris, Wheeler, Pickwick Landing, Guntersville, and Chickamauga Dams were producing hydroelectric power, and another five dams were under construction. Tupelo, Mississippi, was the first town to buy Tennessee Valley Authority power, at 0.7¢ per kilowatt-hour, instead of the 1.7¢ per kilowatt-hour it paid the Mississippi Power Company. In the first six months of Tennessee Valley Authority power in Tupelo, there was an 83 percent increase in consumption of electricity. "It seemed like this old hillside just lit up when we got electricity," remembered a Tennessee Valley farm woman.

Private utility companies, which had been only moderately concerned since they hadn't thought the Tennessee Valley could be profitable, began to pay attention. Threatened by the idea of cheap federal power on a large scale, they plunged into an all-out war on the TVA. The first suit came from thirteen Knoxville ice and coal dealers.[52] By 1936 there were 278 injunctions against the TVA. Nineteen private power companies, cuddled up together in a Wall Street holding company, mounted the most serious suit. Represented by the holding company's president, Wendell L. Wilkie, the suit contended that it was unconstitutional for the government to sell power.[53]

Wilkie, addressing a House Military Affairs Committee hearing, expressed approval for developing the Tennessee Valley but insisted that electricity be sold to the utilities for resale: "To take our market," he said, "is to take our property." After a three-year battle, the court ruled for the TVA. The TVA estimated in 1938 that 41 lawsuits had cost it almost six million dollars. But even though the public was skeptical about private utilities, criticism never entirely subsided, and for decades a comment thought to have originated with Wendell Wilkie was bounced around: "The Tennessee River waters four states and drains the nation."

By now, there were 23 dams on the Tennessee and its tributaries. Nine dams with locks on the Tennessee stair-stepped down the river, dropping barges 515 feet between Knoxville and the Kentucky Dam at Paducah. The TVA dams were centrally controlled to work as a unit, shutting down in case of heavy snow melt and opening up again in drought. When it rained heavily in the Virginia and Tennessee catchments, TVA river-control engineers received rainfall data from hundreds of rainfall stations, reservoir elevation levels and discharge information from each dam, and the flow stages on the Ohio and Mississippi from the Army Corps of Engineers. The gates at the Hiwassee Dam or the Cherokee might then be instructed to close to hold back upstream waters, while engineers at the Chickamauga on the Tennessee might be instructed to release water to make room for the heavy flows moving downstream.

In 1936, the year after the completion of the Norris Dam, Chattanooga was deluged with rain. The dams kept more than a thousand acres dry and prevented the loss of upwards of a million dollars. It poured again in 1937. This time, the saving grace of the TVA was felt all the way to the Ohio. The dams took up to five feet off the cresting waters at Chattanooga, a foot off at Paducah, and six inches off as far down as Cairo, Illinois. "You have to remember," wrote Willson Whitman, "that an inch on the crest of a flood may be as important as an inch on the end of a man's nose."

The TVA distributed electricity to at least part of all the seven basin states at low rates, and many people began to enjoy power for the first time.[54] Social programs kicked in. In the first few years of the TVA alone, the Civilian Conservation Corps planted twenty million trees. Malaria experts undertook mosquito control. Nine thousand miles of virgin shoreline around the reservoirs was opened up to fishermen, boaters, picnickers, and sightseers. One of the Muscle Shoals nitrate plants was refitted to make experimental phosphate and nitrate fertilizers; the results were sold to farmers at cost. Agricultural extension programs taught farmers to save soil by terracing and contour farming and planting soil-enriching crops like clover and alfalfa. Thousands of farmers signed on to a demonstration farm program. Cotton yields rose from 275 to 400 bushels an acre, and corn from 20 to 30.

When war began ripping Europe to pieces in 1939, David Lilienthal, now chairman of the board of directors, saw that America wasn't prepared for its own inevitable involvement. He worried that the United

States was not producing enough aluminum for warplanes and understood that no other place in the country could be geared up to add power capacity quickly. He thought that the TVA wasn't planning boldly enough. "By taking the offensive," he wrote, "by seeking to use deep pulsations of change in these times, can't we build something better than we have known?" Lilienthal began a push for still more dams and steam stations.[55] There were plenty of obstacles. As late as 1942, the War Production Board, claiming that planes and munitions should have priority over power, shut down construction at Fontana. A frustrated Lilienthal again and again struggled to explain the relationship of war production and electricity. "We stand in great danger of losing this war and thereby losing everything because we seem unable to concentrate," wrote Lilienthal.

There were times when survival alone seemed all that Lilienthal could hope for as he fought coal miners and corporations, politicians, the military, and the press. Lilienthal was an able scrapper although it sometimes wore him down. "Up to this hour I haven't once cracked out in publicly displayed anger at nasty newspaper remarks or dirty cracks by my enemies," Lilienthal once groaned privately. "But Godamighty, it's hard to do at times!" Some of the most powerful challenges to the TVA came from inside the government and Lilienthal variously skirmished with the Bureau of Reclamation, the Department of the Interior, the Department of Agriculture, and the Army Corps of Engineers, all of whom competed with one another for jobs and power. "The idea of a regional resources development agency runs counter to the vested interest of existing departments," he wrote. "It is almost incredible how powerful these supposed creatures of a democratic government can be in enforcing their vested jurisdictional prerogatives." During the fights, the TVA agenda evolved and solidified. As Roosevelt often repeated, this was not just about dams but about technical skills used in the public interest.

The politics of war eventually transformed the organization. Under Lilienthal's direction—with Roosevelt's backing—nine new dams and a steam plant were built to fuel war industry. The TVA doubled its power capacity. Muscle Shoals produced 60 percent of the phosphorus used by the U.S. military for tracer bullets, incendiary bombs, and smoke screens, and chemical warfare and aluminum production in the Valley proved crucial to the aircraft industry, just as Lilienthal had predicted. From the TVA-powered Oak Ridge Laboratories came the project's most significant contribution to the war—fissionable uranium 235 to fuel the Manhattan Project. In early 1945, couriers began carrying bomb-grade uranium in special luggage on the 12:50 overnight train from Knoxville to Chicago, and then on to the Santa Fe Chief to Los Alamos in New Mexico.

The day after Franklin Delano Roosevelt died in 1945, a TVA employee drove David Lilienthal to the airport on his way to the funeral and spoke about the president and what he had done for the Valley. "Who are the little people like me going to have to take their part?" he asked Lilienthal. "I spent the best years of my life working at the Appalachian Mills and they didn't even treat us like humans. . . . Sixteen cents an hour was what we got; a fellow can't live on that. . . . No sir, I won't forget what he done for us."

Roosevelt's successor, Harry Truman, called the TVA "just plain common sense hitched up to modern science and good management. And that's about all there is to it." Fifty-eight major and minor dams now controlled the Tennessee River network. No other river system on earth was more thoroughly managed. There were no more disastrous floods in the Tennessee Valley, and while some flood damage continued, it was largely because people kept stubbornly moving into areas still prone to overflow. The TVA pioneered flood-plain zoning, which meant requiring communities to build above flood lines. Eventually, 40 million tons of cargo would be shipped on the Tennessee River each year. Tourism around the lakes became the Valley's third largest source of income, eventually drawing 110 million visitors yearly. Incomes rose 17 percent faster than in the rest of the country.

The TVA improved economic development in an area the size of England, and it did so while caring for its resources—planting trees, stopping soil erosion, and turning exhausted farmland wrecked by nonstop, single-crop depletion into green, productive ground. Millions of acres were restored. Its agricultural program pioneered a range of phosphate fertilizers. People in the Valley used more electricity per household than those throughout the rest of the nation and paid less for it. And, finally, life just plain got better for millions of people in the Tennessee Valley under an agency that had responsibility for both producing power and protecting its people, land, and water. "The one single outstanding achievement of the Roosevelt administration, the single thing that will live for the ages, is the TVA," George Norris said to David Lilienthal. "The time will come, and it won't be very long either, when people will not think of electricity as a private thing at all, and TVA will be responsible for that too."

"The Tennessee River had always been an idle giant and a destructive one," exulted David Lilienthal. "Today its boundless energy works for the people who live in this valley. . . . But all this could have happened in almost any of a thousand other valleys where rivers run from the hills to the sea . . . rivers that in the violence of flood menace the land and the people, then sulk in idleness and drought—rivers all over the world waiting to be controlled by men—the Yangtze, the Ganges, the Ob, the Parana, the Amazon, the Nile."

Americans never tried to replicate the TVA—there would be no political creation so daring, no basin-wide planning administration that would handle responsibility for all land and water decisions, or take on the enemies that such an entity would inevitably create—but its gospel spread fast among countries all over the world, whose leaders came to see it as a model for development. Before there was Israel, Zionists came to see David Lilienthal, looking for a model for the Jordan River. Engineers and politicians came from Europe, the Middle East, Australia, China, Turkey, and India. The Tennessee Valley Authority helped inspire the Cauca River project in Colombia, the giant Akosombo Dam in Ghana, multiple works on the Tigris and Euphrates, the Niger, the Yangtze, and the Snowy River in Australia, among others. The grand idea of the Tennessee Valley Authority, the planned development of a whole region, to Lilienthal's delight, had found its way into the world's thinking.

Lilienthal himself would carry the message to the Indian subcontinent, where a desperate political crisis cried out for a solution and the idea of river basin development was to become the implement of a remarkable international effort to prevent a war.

DAVID LILIENTHAL first visited the Indian subcontinent in February of 1951. At the time, he was only forty-nine and without a permanent job—he had left the TVA to head the Atomic Energy Commission only to resign after the United States began developing a hydrogen bomb—but Harry Truman had another task for him. He asked Lilienthal to become a sort of ambassador for valley development to several emerging nations of the postcolonial world. "The development of these areas had become one of the major elements of our foreign policy," the president later wrote. "In the Mesopotamia Valley alone there could be a revival of the Garden of Eden that would take care of thirty million people and feed all the Near East if it were properly developed. . . . The Zambezi River Valley in Africa and a similar area in southern Brazil could also be converted into sections comparable to the Tennessee Valley . . . if the people of those regions only had access to the 'know how' we possessed."

Although Lilienthal was happy to oblige, he knew that India and Pakistan would present a host of special challenges. Just four years earlier, the subcontinent had been cut loose from the British Raj and partitioned along religious lines into two countries, Muslim Pakistan and overwhelmingly Hindu India. The result was one of the worst tragedies of the twentieth century. At least ten million displaced Hindu, Muslim, and Sikh refugees poured across the new borders. A quarter of a million men, women, and children are thought to have been slaughtered along the way. V. P. Menon, the closest Indian advisor to the last British Viceroy, Lord Louis Mountbatten, wrote of the survivors' lasting bitterness: "The uprooted millions were in a terrible mental state. They had been driven from their homes under conditions of indescribable horror and misery. . . . They had been subjected to terrible indignities. They had witnessed their near and dear ones hacked to pieces before their eyes and their houses ransacked, looted and set on fire by their own neighbors. They had no choice but to seek safety in flight, filled with wrath at what they had seen, and full of anguish for numberless missing kinsmen who were still stranded . . . and for their womenfolk who had been abducted."

Anxiety and anger over water added to the resentment of the people of Pakistan. First, when the Hindu ruler of Kashmir acceded to India, he had taken with him the headwaters of the Indus River system, the source of West Pakistan's water. A dry, largely rainless region, Pakistan could live without Kashmir, but it could not survive without the waters flowing out of its mountains. The Indus is a chilly, turbulent river, 1,800 miles long, carrying twice the flood of the Nile and four times that of the Colorado, flowing out of the glaciers around Mount Kailas in Tibet, one of the most sacred places on earth. To Hindus, it is the abode of the god Shiva; in Buddhist and Jain cosmography, Kailas is Mt. Sumeru, the center of the universe.[56] Fifteen thousand feet above sea level, fed by snow melt and mountain streams, the Indus pushes northwest across the Tibetan Plateau, between the main ridge of the Himalayas and the Ladakh Range into Kashmir, where it turns west through spectacular mountain gorges, then spills onto the plains.[57] Five rivers join their waters with the Indus in what was then West Pakistan—the Chenab, the Jhelum, the Ravi, the Beas, and the Sutlej, which push into the Indus plain like the fingers of an open hand. This is the fertile Punjab— "land of the five rivers"—a mile deep in alluvium washed down from the hills.

Just as worrisome from Pakistan's point of view was that the meandering partition line drawn by Sir Cyril Radcliffe, chairman of the British Partition Commission, placed in Indian hands many of the headworks that controlled the complex system of British-built canals in West Pakistan. It now seems incomprehensible that the flow of water was handled so carelessly, but Radcliffe remained somewhat unrepentant. "Each decision at each point was debatable and formed of necessity under great pressure of time, conditions, and with knowledge that, in any ideal sense, was deficient," he explained several years later in a letter to the author Aloys Michel. "I believe my decision was right, at any rate at the time and for the time, but, of course, the whole experience was unprecedented and terrible."

David Lilienthal admired Jawaharlal Nehru and felt that the U.S. could work with him to find a solution to the problems now confronting India and Pakistan: "India now has a leader," he wrote, "a man we can talk to, can support and help with a clear conscience. We had better try to understand him." Nehru returned the compliment. He, too, was a believer in technology's power to better people's lives. "We dreamt a great dream," Nehru had told his people. "Well, we got political freedom for India but another great problem remained unresolved . . . to ensure that all the people enjoy the benefits of that freedom." To make good on that pledge, to provide electric power and make the new nation self-sufficient in food production, the Indians were already hard at work on three multi-dam projects inspired at least in part by the TVA—Hirakud, Damodhar, and the massive Bhakra-Beas project. Taller than Hoover Dam, the Bhakra Dam on the Sutlej would for many years remain the highest straight gravity dam on earth. Thirteen thousand men and women swarmed over the site each day. "A new India certainly is emerging and at a pace that is remarkable," Lilienthal noted in his diary. "Seeing the building of this vast and most ambitious project—a dam capable of producing more power than Boulder and of irrigating much more land than Coulee—you can feel hope as well as the terrible handicaps in the way this project is being built. . . . Huge Euclid dump-trucks (able to carry more earth than a hundred men), and along with them there are the coolies, the common labor, carrying dirt and aggregate and concrete in little baskets on their heads. . . . Trucks and camels working side by side. Down below us, on the other side of the river, is a long string of grass huts. A few miles away, Bhakra is rising, one of the greatest structures on earth, representing the accumulation of man's technical skills and industrial proficiency." The context may have been disorienting, but the engineers seemed familiar. "Almost every group of technical men among the Indians we met includes several who have visited TVA or trained there," Lilienthal added. "It is like meeting old friends in a strange setting."

But Lilienthal also saw that all of India's potential gains—and Pakistan's very survival—depended on somehow forestalling warfare over the waters of the Indus. In Karachi, Pakistani Prime Minister Liaquat Ali Khan told him: "It is a great tragedy that these two countries that are so close to each other in so many ways, so dependent upon each other, should be trying to destroy each other." And in Lahore he witnessed the Pakistani public's fury when India temporarily halted canal flows in order to perform maintenance at the headworks. "No army with bombs and shellfire," he wrote, "could devastate a land as thoroughly as Pakistan could be devastated by the simple expedient of India's permanently shutting off the sources of water that keep the fields and the people of Pakistan alive."

Neither side seemed willing to back down. "India and Pakistan are today on the very razor's edge of a war that would directly involve more than three hundred and sixty million people, one sixth of the world's population," Lilienthal wrote in *Collier's* as soon as he got home, "and might well set fire to the whole Moslem world from the Arabian Sea to the Valley of the Nile. Because the United States has already deeply committed itself in this controversy, the outbreak of war would undoubtedly put the United States of America into another and even bigger 'Korea.' "[58]

To forestall disaster, Lilienthal then made a suggestion that few other men could soundly have offered: engineering, he said, could provide the answer. He proposed joint development of the Indus system, to be built and operated by both countries with international help. "These people will live or die by how they handle the waters of the Indus River," he wrote. "Properly handled, the river has plenty of water for everyone.

With a mutual project under way, with these brothers again working together on these things, the political issue of Kashmir may be solvable, the UN's heavy commitment discharged without force, and 'another Korea' prevented."

Lilienthal personally delivered a copy of the magazine containing his article to the Pakistani ambassador in Washington on July 24, 1951. The next day, he received word that Pakistan would accept the proposal as it stood. Seven days later, Eugene Black, president of the World Bank, called Lilienthal: "Wonderful idea, absolutely sound, makes good sense all around. Fascinated by the idea. Now, how do we go about getting it adopted?" The World Bank, which had been withholding loans to either country until progress on the water dispute had been made, informed both prime ministers that the bank would like to mediate a settlement. India and Pakistan accepted the bank's offer.

Negotiations ground on for two years with Pakistan clinging to its historic rights to the rivers and India insisting that it held territorial sovereignty. A cartoon of Nehru appeared in *Punch* with the caption: "He believes in the principle of non-attachment, except in the case of the Indus Catchment."[59] In February of 1954, the World Bank proposed dividing the waters in half: the waters of three western rivers—the Indus, Jhelum, and Chenab—were to go to Pakistan, and those of the three eastern—the Ravi, Sutlej, and Beas—to India. Pakistan countered that there was no provision for reservoirs and without them the western rivers were inadequate to meet even their present needs. "I have been around these areas which are going to be affected by the withdrawal of waters by India," the new Pakistani prime minister Ayub Khan told Eugene Black. "People have told me very plainly that if they have to die through thirst and hunger they would prefer to die in battle and they expected me to give them that chance. . . . So this country is on the point of blowing up if you don't lend a helping hand. This is a human problem of a grave nature and cannot be blinked away."

With his own representatives he was more cautious: "Gentlemen, this problem is of far-reaching consequences to us. Let me tell you that every factor is against Pakistan. I am not saying that we should surrender our rights but, at the same time, I will say this: that if we can get a solution which we can live with, we shall be very foolish not to accept it." A panel of Pakistani engineers proposed building two large dams, three smaller dams, and a series of link canals at a cost of $1.12 billion. India, which didn't want to pay for those dams, protested. Black asked Pakistan for a more realistic budget and proposed that India's portion of payment be a fixed sum. He also offered a $56 million World Bank loan for an Indian dam on the Beas River. Pakistan would receive $541 million in grants from "friendly" Western governments, $150 million in loans from the U.S. and the World Bank, and $174 million from India.[60]

It took nine years of hard work in Delhi and Islamabad and Washington, but the Indus Water Treaty, first proposed by David Lilienthal in 1951, was finally signed on September 19, 1960. The waters of the three eastern rivers went to India, and the three western to Pakistan. India promised a ten-year grace period during which water would be supplied to its neighbor while dams and canals were under construction.[61] David Lilienthal was gratified to have helped achieve a peaceful agreement, but privately he felt that King Solomon had cut the baby in half. His dream that Pakistan and India would work together on an engineering system that would manage the river basin as a single entity was dead.

Still, through an elaborate and detailed treatment of most aspects of water use in the Indus Basin, and backed by financial support for a bold system of engineering works, the treaty gave India and Pakistan a way to live next door to each other in peace. Relations between India and Pakistan have not improved since the Treaty was signed, but India never held back its payments to Pakistan's Development Fund. Even when waging war upon each other, neither nation has ever attacked the other's dams or canals.

THE SIGNING OF the Treaty in 1960 was the signal for Pakistan to undertake an all-out effort to build a monumental array of dams and canals during the ten-year grace period granted by India. "This was like a war," remembers Syed S. Kirmani, a squarish, pleasant-looking man in his seventies who had been a member of the engineering team that had reported to Ayub Khan. "These were huge works. Some of the canals were bigger than the Thames. It was the first time that canals of those capacities were built. What would happen—well, nobody had a very clear idea. Everybody was after us. They said we had sold the rivers, that we were traitors to our country. We had a very, very bad time. It was difficult to show our faces." They had little time to feel sorry for themselves. It was time to build.

Kirmani was an accidental Pakistani. A Muslim from the city of Madras in the south of India, he had been working in Lahore at the time of Partition, so he and his family remained behind in Pakistan. I spoke with him in the offices of the World Bank in Washington, where he has since built a distinguished career. Although a principal architect of the engineering that became the Indus Basin Plan, in 1960 he was still relatively junior in status, working out of a closet-sized office in Lahore. Kirmani was completely unprepared when he was asked to oversee the entire engineering assault on the Indus Basin.

"I called the project managers when WAPDA [Pakistan's newly formed Water and Power Development Authority] asked me to take over, and said, 'I'm very junior to you, if any one of you wants to take my place, you are most welcome,'" Kirmani remembers wryly. "But *nobody* wanted it." Chief Engineer Syed Kirmani now had to oversee building on an unheard-of scale. According to the treaty, four hundred miles of large link canals, which would carry water from the western rivers to land formerly supplied from the eastern rivers, were to be finished within five years. All other canals were to be finished within ten, with overtime penalties to be paid to India. "In those days the feelings of Pakistan vis-à-vis India were such that people would say, 'We would die rather than pay a penalty to India,'" says Kirmani. "We knew we would be lynched if we failed."

There were precious few resources to draw upon. Pakistan had no cement factories, no aggregate stone, and few working quarries. The port of Karachi was small and crowded, so an extension had to be built to accommodate supply ships. Locomotives and railroad cars had to be imported. "We didn't have engineers either," Kirmani laughed, "so we went to the colleges and universities. Before engineering students completed their final year, we told them, 'If you pass you will continue and if you fail you will still continue.' We went after retired persons—everybody who was able to walk."

Foreign contractors took on different parts of the massive project, sometimes developing new technologies as they went. Digging the deep Qadirabad-Baloki Link canal, for example, the American firm Morrison-Knudson designed excavators with buckets so big a truck could sit comfortably inside them. "On the sixty-mile Cheshma-Jhelum Link Canal," Kirmani recalled, "the quantity of earth to be moved was almost 110 million cubic yards, so we used a bucket-wheel excavator, a huge thing, before this only used to mine coal. Our consultants said it wouldn't work, but I went to Germany and became convinced otherwise. It was amazing how prices came down with this equipment. With the first canals we got quotations of $1.10 per cubic yard, but by the time it came to the bucket-wheel excavator it was 77¢ per cubic yard."[62]

The Pakistanis built all of the access roads, railroads, and housing on each project, so that when foreign contractors arrived they were ready to go. On one of the smaller link canals, a local contractor named Mir Aslam Khan was hired, to the horror of consultants, who came shouting to Kirmani that the man was excavating with donkeys and baskets. "We were in a fix," Kirmani told me. "We faced not only a technical problem but a political problem, because he was the lowest bidder and met all of the specifications. But the consultant said, 'Look here. How can he compact? We want sheepsfoot rollers! These donkeys can't do that.' But we kept him on. We used to call the big earthmovers D-5, D-7, D-9, and so on. So we called the donkeys D-1. And D-1 worked well, it didn't require spare parts. In fact, it had excess spare parts. We told the consultants, 'Your job is to check the density of compaction. If it is not satisfactory, let us know.' That fellow used old stonerollers. He put three pairs of bullocks on a roller and they ran slow and steady while the donkeys brought the earth in small heaps. The work went well; he completed the job ahead of schedule, got bonuses, bought a Mercedes, and went around like a lord."

In the course of building, Kirmani supervised construction of two of the largest dams on earth—Mangla on the Jhelum, built by a consortium of eight U.S. firms, and the billion-dollar Tarbela on the main stem of the Indus, which remains Pakistan's largest source of electric power. Mangla is an especially striking piece of work, contoured around a hill topped by a sixteenth-century fort. Its spillway drops in long, graceful waves to the plain below, concrete slopes that absorb the energy of the high drops of water and prevent the riverbed from being dug up and washed downstream. Unusual care was given to the watershed at Mangla, in line with the ecological care exercised in the Tennessee Valley. Foresters, civil engineers, and agronomists worked with farmers, building silt traps to reclaim land and check dams laced with vegetation to catch runoff, and planting trees in the watershed to stop erosion and silting. As a result, silt loads dropped dramatically: in just one area, in one ten-year period, thirty-three tons of sediment per acre-foot of runoff were reduced to seven.

Kirmani and his team made their deadline. Pakistan, supported by the World Bank and a handful of Western countries, had pulled off a series of staggering engineering feats: the world's largest contiguous surface irrigation system; two of the largest dams in the world; nineteen barrages and headworks; forty-three main and twelve link canals; a total of 36,000 miles of canal distribution and 89,000 local watercourses—over a million miles in all. The Indus and its canals now irrigate more than 35 million acres of land. "When you are cornered and you have very few options, you try to stand up to it," says Kirmani. "Our people worked with such enthusiasm and commitment. I think it was the times, the challenge—the determination to survive."

———————

ACROSS THE BORDER in India's Punjab, on October 22, 1963, Prime Minister Nehru dedicated the Bhakra Dam. He called it "the temple of a free India, at which I worship." The building of that temple, the keystone of the Bhakra-Beas system of dams and canals that was to distribute India's share of the Indus waters, had also been a monumental feat.

It was for many years the highest straight gravity dam in the world. And although its reinforced concrete surface is stained from thirty years of heat and monsoon, it remains magisterial. The water, which pounds down the river after having done its work in the generating plants, is tourmaline blue. Bhakra's power lines and canals march downstream for 225 miles through the Punjab, through Haryana State, and on to Delhi.

The visitor's center features wall-sized black-and-white photographs of Nehru and the two men most responsible for getting the job done: India's chief engineer, A. N. Khosla, and Harvey M. Slocum, the American construction engineer hired to oversee the Indian firm of Punjab Engineers Pvt. Ltd., which did the work. The bespectacled Khosla is immensely dignified in the presence of his handsome prime minister. But Slocum, who had worked on the Grand Coulee Dam, is clearly an American import—craggy, canyon-smart, and not intimidated by anybody, not even Jawaharlal Nehru.

Slocum was lauded in the Indian newspapers as a "rugged, nononsense technician who would refuse to be balked by obstacles." My father-in-law, Champ Ward, remembers sitting at the next table in a Delhi restaurant in 1957 and overhearing Slocum proclaim: "I told them I could take that river up over the mountain!" One Indian reporter wrote, "Mr. Slocum, blustering and bullying his way through Himalayan barriers . . . drove his men mercilessly and Mr. Nehru was at hand to prevent discouragement by endlessly repeating the ultimate advantages which would flow." A Slocum one-liner—"They talk dams. I build 'em"—was recently reported to me in the Punjab as if he had said it just a month before.

There are also taped interviews with workmen at the visitor's center. "We worked under great difficulty," reports one man cheerfully. "Our hands and feet would disappear in the cement. But we felt it was our duty and we had to do it." "Nehru told us what would happen," reports another proudly. "We didn't know what electricity was. Before Bhakra we couldn't even build a bicycle."

Harvey Slocum said that Bhakra was the toughest assignment of his career. It took 15 years, 13,000 laborers, 300 engineers, and 50 foreign experts to bring it into being, but it brought almost 4 million new acres of irrigated land into production, reduced flooding, expanded the country's appetite for water control, and became the first component of a series of interrelated water works that use Himalayan river water over and over again. Before the Indus Treaty, the Indian Punjab produced just 3.5 million tons of grain a year; that number now stands at better than 21 million tons.[63] Today, every village in the Punjab has electricity. In a power-poor country that too often relies on burning coal to provide electricity, the Bhakra-Nangal system is a vital source of power with an installed capacity of about three thousand megawatts (MW), enough for 3 million people. It has paid for itself many times over.[64]

In New Delhi, I talked about Nehru's visionary dam with the late founding chairman of India's Central Ground Water Board, B. B. Vohra. "Electricity plus water—a great success," said Vohra. "The dam has controlled the frequent floods in the Sutlej and the hydropower it produces laid the foundation for progress in the Punjab. Land that is now developed used to be entirely under wild grass—a den of thieves. Bhakra is one of the few big projects that is a great success story."

In both Pakistan and India, the engineering works of the Indus system are still proudly described as being based on the TVA, and some of its achievements were successfully replicated in both countries—abundant affordable electricity, flood control, and water for irrigation. But, as David Lilienthal was one of the first to point out, the Indus Treaty failed to achieve the greater goal of the Tennessee Valley Authority, which was to combine the development of land and water with its conservation across the natural geographical boundaries of a river basin for the benefit of all the people who live there.

Over the years, both India and Pakistan would stray further from the TVA's example, abandoning environmental and social concerns in the headlong rush to modernize at any cost. We should not be surprised. After all, the Tennessee Valley Authority itself didn't long outlast the people whose ideas inspired it.

THE TVA LAKES still stretch along the valley between the rolling hills—Normandy, Pickwick, Chickamauga, Nickajack, Watts Bar, Hiwassee, Chatuge, Norris, Boone, Kentucky, and Cherokee. Flying over the valley at night, the bright lights dancing along the river make a necklace of dams, dozens of them, in all sizes and shapes. From the air, everything still seems as it was in David Lilienthal's time, but appearances are deceptive. On the ground, nearly everything has changed.

S. David Freeman saw it all happen. The first Tennesseean ever to head the TVA, he has known the region since boyhood and witnessed the evolution of opinion among its people. A tall, gray-haired man in an eye-busting red shirt, jeans, and cowboy hat, he seemed cheerfully out of place in his genteel New York City high-rise. "I still remember the family that lived upstairs in '48 when I was just married and living in a Knoxville apartment," he said. "I'd sit and talk to the grandmother, Mrs. Runyon, who was from the mountains. She said to me one day: 'David, if them bums out in Oak Ridge don't blow up and kill us all, them dams will burst and flood us.' She thought that if the TVA had just left her mountains and her people alone she'd be better off. But by now the grand plan of the TVA has faded into the woodwork and not many remember what it was like before the dams came."

David Freeman loves the ideas that were once at the heart of the TVA. "The concept that you take a river valley and maximize all the potential benefits and funnel those benefits to the people in the valley was powerful. I haven't visited any part of this planet that hasn't heard of the TVA. I was one of the first visitors to Fujian province in China and was told that the Min River Authority there was patterned after the

TVA. I went to Pakistan and signed an agreement with the Water and Power Development Authority to exchange engineers with the TVA. The TVA reached a worldwide audience. It was a gigantic success."

World War II ushered in the first changes, commandeering resources from social and environmental efforts for military purposes. Several years after the war ended, half of the TVA's production was still being consumed by the Atomic Energy Commission facilities at Oak Ridge, Tennessee, and Paducah, Kentucky. The election of Republican Dwight Eisenhower as president in 1952 accelerated the process. Providing the cheapest possible power took precedence over the social and ecological concerns that had once characterized the TVA. Seven coal-burning plants were built during the Eisenhower years. After a time, 80 percent of the TVA's power was being generated by coal, and the TVA had become the nation's largest sulfur dioxide polluter and its largest buyer of strip-mined coal. It resisted pressure to install anti–air pollution devices and opposed improved health-and-safety controls in coal mines.

"Keeping electric rates low became a kind of religious thing," Freeman grimaces. "The TVA, which had put green cover on the mountains by teaching people how to control soil erosion, was now ripping up the country with strip mining. It was a sad sight to see. I remember, when I was still just a TVA lawyer, I asked, 'Why in the hell can't we put a clause in the contract requiring the coal company to reclaim the land?' The argument from the power people was, 'That will raise the price of coal and the price of electricity, and our job is to sell electricity cheaply.' Instead of putting a green cover on the soil, they were tearing it up in the name of progress."

Progress also brought nuclear power to the Valley. The TVA became the largest producer of nuclear-generated electricity in America[65] and saddled itself with tens of billions of dollars of debt, under which it still staggers. Over a thirteen-month period in the mid-seventies there were 65 "abnormal occurrences" in the Brown's Ferry facility, the world's largest nuclear generating plant, incidents that triggered national debate over the safety of nuclear plants.

In 1977, David Freeman, who had worked in Washington during the Kennedy administration, was sent back to the Valley by President Jimmy Carter. "Carter thought the outfit had become just another utility," remembers Freeman, "so he sent me down there to try to bring it back to its roots. But it was forty years later and times were different. If I spent one nickel of the power money on something else, the people and politicians of the Valley were against it."

Freeman threw himself into giving the Tennessee Valley Authority a new lease on life.[66] "With the help of my colleague Richard Freeman, we re-created a national purpose for the TVA," he tells me. "We launched about the largest energy-efficiency program in the nation. We raised the alarm about acid rain and helped get legislation through to stop it. Working with the Department of Agriculture in West Tennessee, where land was eroding, we got no-till farming practices installed. That led to amendments, which I think have done a lot to stop soil erosion. And we worked hard on economic development."

The team also undertook a billion-dollar program to clean up sulfur dioxide emissions by installing scrubbers and using lower sulfur coal. They planted trees, installed ten thousand solar water-heaters in Memphis, worked on preservation of water quality, adopted a "no more dams" policy, even experimented with solar ponds and wind power. "We reached one million homes with energy conservation," Freeman recalled, "and in so doing built the equivalent of a large nuclear power plant at a fraction of the cost. It was a very successful program and we could have set the model for the nation. I shut down eight nuclear plants. Although when it was all over you still had an electric-power dog and a tail of non-power programs."

David Freeman left the government in 1984. "I hate to say this, but I feel like I was a heart transplant to the TVA," he grins ruefully. "Shortly after I left, I was rejected. The nuclear plants that we shut down stayed shut down, so we made some lasting changes, but we were not able to reinstill TVA's former vigor in a permanent way because the political process has gone in the other direction. Not long after I left, the non-power programs were virtually eliminated by the new appointees."

Only eleven percent of the electricity now generated by the TVA comes from hydropower. Twenty-four percent is nuclear. Sixty-five percent is coal. "Maybe it's time to retire those revered letters, TVA, and call the outfit the Tennessee Electric and Power Company and forget it."

"I think the lesson is that old water agencies never die, they just fade away," Freeman continues. "That's not necessarily bad. Their mission was to harness the rivers, to build the dams in a grand plan. You can only do that job once, then it becomes a caretaker job. What is lacking is a new, exciting mission."[67]

Freeman believes the TVA could be salvaged and serve as an example for all those places in the world that still have water resource opportunities. "The TVA still has under one umbrella as much talent in environment and energy as anywhere in the world," he insists. "It ought to be the place where we hammer out solutions to development in harmony with the environment. It puts a power system to use in a problem-solving way, a job the TVA is uniquely capable of doing."

Except that the residents of the Valley aren't interested. Their valley is developed. Everybody has electricity. They want to go about their business and pay as little as possible for power. But Freeman urges me not to lament the passing of the big national agencies, and instead to remind people of what a difference they made. "The TVA is a success because the idea is so powerful," he stresses.

The revolutionary idea behind the TVA, David Lilienthal wrote, was that one agency was entrusted with responsibility for everything, that no one activity could be considered an end in itself, and the unified development of resources must be the common purpose. Systemwide planning is the TVA's greatest legacy. It's more important than ever, Freeman points out, in poor countries where there are too many people for the water and the land. "Once you learn the lessons that we have learned and apply them," he says, "multipurpose analysis is more relevant than ever."

"All these men and tools were never quite enough, of course, to do everything that was expected of them," the late environmental historian T. H. Watkins wrote of the TVA. "But for all its failings, it should be remembered that this kind of thinking got things done in a big way—never perfectly, sometimes badly, but almost always with a dimension of hope that is breathtaking to observe from the distance of more than fifty years."

That dimension of hope—the promise of enormous good that can be done for all the people of a river basin—remains relevant in our time of scarcity and overcrowding. "A fundamental change in resource development . . . must begin at the beginning," wrote David Lilienthal. "In the minds of men, in the way men think and, so thinking, act. And what is true of our region is, I deeply believe, equally true of regions and people everywhere."

Chapter 4

DRY, DRIER, DRIEST

Greening the Desert

*Water is the true wealth in a dry land; without it,
land is worthless or nearly so. And if you control the water,
you control the land that depends on it.*

WALLACE STEGNER, *BEYOND THE HUNDREDTH MERIDIAN*, 1954

Beryl Churchill is a fine-looking woman—tall, slim, and gray-haired. As a farmer who has worked hard all her life she feels baffled, after years of honest labor, that there are people who find her use of irrigation in the West reprehensible. "Out here, water is the magic wand," Beryl told me, her jaw set firmly. "But I don't think we waste water."

Beryl and her husband Winston farm nearly eight hundred acres of sugar beets, malt barley, beans, broccoli, and cabbage in Powell, Wyoming, courtesy of the Buffalo Bill Dam and Shoshone Irrigation Project. Winston's father farmed this ground, as did his grandfather. One of the Churchills' sons now works it with them. The other two run farms and a ranch nearby.

Beryl grew up in Cody, just west of Powell, the daughter of a contractor and the grandniece of Martha Jane Cannary, better known as Calamity Jane. She learned to fly a plane at sixteen and used to help set up camp in the mountains for her grandmother, one of America's first female hunting-and-fishing guides. That Beryl is spirited and strong-minded is a product of both family and the wild country around her. When she decided to marry Winston, her parents were concerned about her taking on the hard life of a farmer's wife. They needn't have worried. I've rarely met a happier woman.

The Churchills have done well for themselves. They paid for their air-conditioned tractor by writing a check, as they did for four smaller tractors, a fancy combine, and the trucks. Their snug, white house is not all that large, but it has a fireplace, picture windows overlooking fields, and a kitchen—almost bigger than my apartment—where the living happens, where business is transacted and neighbors come to chat. The family and their life appear so attractive that it is easy to forget how long and strenuous their days are.

Powell, a plain town of 5,770 people, sits on flat land along the Garland Canal in the Bighorn Basin and is as tidy as a church lady's living room—a pleasant, treed settlement whose schools and shops sprang up out of nothing to meet the basic needs of its homesteaders. It's the kind of town where a would-be bank robber in a ski mask walked into the First Federal Savings Bank a few years ago demanding that the teller hand over all her money, she refused with an indignant "No," and the bandit ran like the wind. There's a small college in Powell and a new Budweiser plant outside of town that buys the farmers' malt barley. Down the road in front of the Western Sugar Mill sit stacks of sugar beets, the size of coal piles at an electric plant. Powell, named after the grandfather of Western water knowledge, John Wesley Powell, is a homey town with no signs of decay. Powell works.

But pretty Powell, Wyoming, is not an easy place to be a farmer, although Beryl and Winston Churchill would never say so. This is a desert in a mountain valley. Winters come early, stay late, and are unsparing in their duration. Winds are ferocious and the growing season short. There is no rain worth mentioning, barely six inches a year. The ground water isn't reliable—most is brackish or contains sulfides. The water in the rivers, then, is Wyoming "white gold"—snowmelt from the mountains—a precious currency pouring out of mountain ranges in rivers and streams.

Agriculture is the biggest user of water everywhere in the world. Eighty to ninety percent of all consumed water goes onto fields, and commonly only half of that touches crops. We've learned a lot in the past hundred years about how to make irrigation work. But this knowledge has been slow to make its way into the fields, especially in poor countries, and the record of irrigation projects is commonly a treacherous read. Most failures occur because projects are constructed without taking human or ecological effects into account. A Dutch engineer once commented, "An irrigation canal is not a technical matter but a social event." Yet systems are all too often planned without consulting farmers or studying local systems. Engineers from green, rainy countries have put unsuitable systems in dry, thorny ones, leaving decaying, unworkable canals and heavy debt loads. Health officials in tropical places have not been included in planning irrigation schemes to prevent waterborne diseases like schistosomiasis. Not enough water has been left in rivers for the survival of ecosystems. Worst of all, governments and builders, in place for the short term, justify projects for short-term gains and don't spend the money necessary to make irrigation work over the long haul.

I'd been reading a lot about the problems and perils of irrigation before I flew to Wyoming to meet the Churchills. Some of the literature on irrigation farming, particularly the tales of water overuse in California's Imperial Valley were so horrifying that I couldn't reconcile them with the informational books from the World Bank. I wanted to know if there were places where irrigation actually worked or if greening the desert was a fiction.

Within fifteen minutes of my first meeting with Beryl, we walked out in a field to heave a canvas dam out of a ditch, allowing the water into her fields. She'd requested the water two days before by placing an order card in a roadside box where a ditch rider—in a truck now, not on horseback as he would have been in earlier days—picked it up and later moved water from district canals into the Churchill ditches. The Churchills are careful farmers; they know exactly what goes into and comes out of their ground. Like other farmers in this district of the Shoshone Project, they have soil tests taken to indicate the amount of potassium or nitrates in the soil and what fertilizer is needed. They try to avoid pesticides. They know every detail of the irrigation system and at any given moment can tell you how much water is in the reservoir behind Buffalo Bill Dam.[68] They maintain a ditch to pick up excess water and send it back into the Shoshone Irrigation system for reuse.

There are more than 600 full-time farmers and 300 part-time ones on the 35,000 acres of Garland District, one of four districts on the Shoshone Project. The average farmer works two hundred acres. Only a handful own more. "People like Winston's grandfather set the stage for us," Beryl told me. "They worked hard to get it established, and made the payments. We are fairly secure because they built us a good base. The struggle is over. We can experiment now; change the irrigation system or try a new crop."

It's hard to imagine how the first homesteaders survived their early years here. Beryl has found letters and diaries of settlers' wives who arrived in the early 1900s and wanted to turn around and go back the very same day. "It seemed like the end of civilization," wrote one woman. "I was homesick so many times. I felt like I was in a jail, with all these many mountains around me. I drove horses and plowed in the fields, but I was scared to death." Early photographs, made before the water came, frame dismal, flat land, empty of anything except shanties put up to stake a claim. To get an idea of what life used to be all across the Garland flats, you can drive just a few miles north to Polecat Bench where there is still no irrigation or anything else. It's sobering after wandering through the green districts around the canals to see so much cracked and barren earth.

Beryl and Winston Churchill farm their land courtesy of an idea first put forward by Colonel William F. Cody, better known as Buffalo Bill. At the end of the nineteenth century, Cody foresaw that the use of mountain water would change the land, and distributed an advertise-ment: "Fine land wants the plow. Unlimited water wants the cultivator." These beckoning words accompanied a drawing of Buffalo Bill himself, standing hirsute and confident at the headgates of an irrigation canal. The reader was assured, "Price reasonable and in small payments." There wasn't a bigger celebrity in the United States in 1900 than this hero of the Old West—frontiersman, Indian scout, showman, and would-be entrepreneur with big-time dreams for the New West. "You know I am a broad gazer," he wrote his sister Julia.

Cody had led military expeditions into Wyoming's Bighorn Basin in 1874 and 1876 and knew the land well. In 1888, he returned to stay. Three big snow-fed rivers flow out of these mountain ranges, but Cody was interested in just one—the Shoshone, called "Stinking Water" by the Shoshone Indians because of the foul-smelling mineral springs and gey-sers along its edges. In the windblown Wyoming lowlands, through which the Shoshone moved, Buffalo Bill glimpsed a rich future.[69] The soil was good and the sloping gradient of the land ideal for irrigation ditches, and Cody envisioned that a million acres of sagebrush flats in the Bighorn Basin would become "the new Eldorado of the West." Cody bought a ranch in the basin and moved cattle onto it. Just below a spec-tacular thousand-foot canyon on the Shoshone River, he and a few friends laid out a town and named it after him. They built a post office and a general store, and helped to convince the Burlington and Missouri Railroad to bring in a rail line. Soon a thousand people had made Cody their home.

Buffalo Bill Cody knew that without water—water for irrigation, for electricity and manufacturing—there would be no development in Wyoming. In the 1890s Cody's citizens already paid twenty-five cents for every barrel of water hauled up from the river. In 1897 Cody formed the Shoshone Land and Irrigation Company, claimed 400,000 acres of good land in the basin, and began building 150 miles of canal to water it—the Cody Canal. "When I die," he told a reporter, "I want the peo-ple of Wyoming who are living on the land that has been made fertile by my work and expenditure to remember me. I would like people to say, 'This is the man who opened up Wyoming to the best of civilization.' " In the end, even Buffalo Bill couldn't raise the enormous amount of money needed to irrigate the Bighorn Basin. By 1896 only a dozen farmers were receiving water, and in 1902 Cody sold the land to the federal government.

At the turn of the twentieth century, nine out of ten private irrigation companies were failing or close to bankruptcy, and settlers looked to the federal government for help.[70] Cody and his cause had a friend in President Theodore Roosevelt, already a fervent proponent of irrigation as a means to aid westward expansion. "When the works are constructed to utilize the waters now wasted," Roosevelt remarked in a 1903 address, "happy and prosperous homes will flourish where twenty years ago it would have seemed impossible that a man could live." Roosevelt's irrigation gospel was preached to an assenting choir. Waters not used were waters wasted. A desert was meant to bloom. The 1902 Reclamation Act authorized federal planning and construction of irrigation works in sixteen western states and territories, and created a revolving fund out of the sale of public lands to pay for projects. Water users would pay back the actual costs of building dams and canals. It was a perfect time to build along the Shoshone River.

One reason for the failure of Bill Cody's plans was that he hadn't gazed broadly enough. He hadn't found a way to store the Shoshone's winter run-off for use when the river was low. Government engineers would enlarge his vision. Between Rattlesnake and Cedar mountains, in a narrow granite canyon ten miles west of Cody, they built a dam higher and more daunting than any that had been built before, the first to be made entirely of concrete. The Shoshone Dam, later renamed after Buffalo Bill, could store enough snowmelt to outlast the dry summers of the Bighorn Valley. The land running east from the Buffalo Bill Dam dropped about twelve feet each mile so that the water in the Reclamation Service Canals poured wonderfully well into the good alluvial soil on the flats—sandy and clay loams with enough rock and gravel underneath so that it would drain naturally.[71]

Winston Churchill's grandfather, Frederick Griswold Churchill of Weathersville, Connecticut, was looking for a more benevolent climate for his hay fever and asthma when he read about the Shoshone Project in a farmers' magazine. He came to Powell, Wyoming, in 1909, shortly after the Department of the Interior opened up basin land to homesteaders. The Garland Division of the Shoshone Project watered almost ten thousand acres on thirty-nine farms by the time Frederick Churchill arrived. Churchill was a farmer as were most of his fellow homesteaders, but his new neighbors included bankers, druggists, doctors, and teachers starting new lives. They came from all over the United States, and almost none of them knew anything about dry-land irrigation farming or the dangers of overwatering.

Even before the Shoshone Dam was up, several hundred farmers had dumped so much river water onto the land that the canals began to leak, water tables rose, and the land began to bog. "Water was used in excess quantity and wasted shamefully," wrote Thomas N. Means and H. N. Savage of the U.S. Reclamation Service. "Crop yields have been reduced, the rise of alkali has been started, cellars have been flooded, crops have been cut by hand or abandoned on account of ground being too wet, and there is now prospect of serious and permanent damage to such valuable land. . . . New farmers fresh from the rain belt cannot be taught irrigation in a year or two."

The government moved in, bringing a forty-ton horse-drawn dragline, nicknamed Vulcan, and dug drains every quarter mile to carry off excess water. A steam shovel succeeded the Vulcan and things improved quickly. Government agents preached caution to the Garland farmers who set about learning how to direct water, about proper slopes and leveling land, and the land started producing again.[72] By 1914, nearly nineteen thousand acres were settled. Fields were full of wheat, alfalfa, hay, oats, potatoes, and sugar beets.

"All my interests are with the West," said Buffalo Bill finally. "The modern West. I have a number of homes there, the one I love best being in the wonderful Big Horn Valley, which I hope one day to see one of the garden spots of the world." It may not have become a garden spot, but the well-watered irrigation districts of the Shoshone attracted a solid, hardworking lot of settlers. "Wyoming has been settled by a superior class of people," boasted a government pamphlet. In his eagerness to irrigate the northwestern corner of Wyoming, Buffalo Bill Cody, a man out of the Old West, had plunged himself into the struggle for a newer West, a fight that could only be won with water.

While staying in Powell, I got used to having my breakfast in the local diner, alongside robust farmers in baseball caps and colorful suspenders who moved from table to table, talking about crops and markets and what was going on in Washington. "Where's Henry?" queried one man. "He's stuck out in his sugar beets," shouted another and the rest smiled and nodded their heads knowingly. "Have a good day," a waitress called to another farmer heading toward a shiny new pickup truck. "Well, hell, I will." He grinned. Nobody seemed to doubt that he would.

"My son Todd and his wife, Janice, over on the Two Dot Ranch, are crying for moisture," Beryl Churchill told me in a letter that described a typical season in western Wyoming. "Jan says the only rain they are getting is something like someone spitting then leaving." The Churchills are militantly involved with water politics. "If you don't control water out here," says Beryl's husband, "it's over. It's finished. You're dead!" They have both put in years on the board of Garland Irrigation District and no major project has been undertaken here without one of them playing a role. Beryl was a commissioner of the Wyoming Water Development Program for eight years. "This irrigation project is one of the most successful small projects in the West," she told me. "This is a progressive project, it always has been."

Irrigation demands foresight and attentiveness. Let that concern slip and ditches go to hell, water and salts build up, and land bogs. The Gar-

land Irrigation District around Powell is successful for the only real reason that any irrigation project is ever successful: good local leadership
and the involvement of all of its farmers.[73]

Although these districts have done everything possible to manage
costs and land, maintaining a large irrigation system requires big money.
Between 1977 and 1986, the Shoshone Project borrowed $19 million
from the Bureau of Reclamation to change surface ditches to underground pipes. Watermaster Dean House says that the pipes control seepage, reduce maintenance, and improve deliveries. "Pipes are faster," he
said. "In a dirt ditch you'd put the water in and maybe six hours later it'd
come along. In a pipeline or even a cement ditch, in an hour it's taken
care of. Pipe also means that you can keep a constant head, you don't
need to worry about weed control, and in storms, snow won't plug them
up." Two of the four districts, alarmed at the high cost of pipe, took the
money scheduled for one year's work and built a concrete plant to make
their own. Commercial manufacturers scoffed, but by the time Garland
and Heart Mountain were through with their program, they were selling
pipe to other districts. The plant has paid for itself several times over.

I asked what would have happened if they hadn't been able to borrow
the rehab money. "Our operation and maintenance costs would have
been raised substantially," House told me. "We would have lost more
land to seepage, and production would have dropped off. Maybe we'd
have lost crops because of washouts, because some of those structures
were in bad shape." "Operation and maintenance would really have gone
up," Beryl added. "I believe that by rehabilitating the system, we have
fewer environmental problems. But that's a farmer talking."[74]

The irrigators also built the first privately owned, low-head
hydroplant in the United States. This small-scale plant takes advantage
of water coming down the canal, a small drop—head refers to the size
of the drop—that doesn't look like much but generates 2.9 megawatts of electricity yearly, which they sell to Pacific Power and Light. It's
a rule of thumb that capacity of one megawatt serves roughly about a

thousand people for a year so the tiny plant on a canal drop produces enough power for about half the houses in Powell. Commissioned in 1983, the plant had paid for itself by 1997 and now brings in $350,000 a year in revenues, which are plowed back into the water district.

Beryl Churchill thinks a lot about water. Because her family and her neighbors have worked so hard to manage theirs, she is angered by the anti–dam and irrigation sentiment of some environmentalists. She is particularly pained by lines like these from Marc Reisner's popular *Cadillac Desert*: "The Bureau of Reclamation set out to help the small farmers of the West but ended up making a lot of rich farmers even wealthier at the small farmer's expense. . . . We set out to make the future of the American West secure; what we really did was make ourselves rich and our descendants insecure."[75] Beryl is adamant that the way in which dams and Western farmers get lumped together is unfair. "There are different conditions," she says. "You can't just put all farms in one box." She freely admits that farming affects the land: "Anytime grassland and desert is changed into farmland, and anytime irrigation water flows through a field, the land is changed. Forever." But the Shoshone farmers grow one crop a year and can't afford to have the ground bogged or their water poisoned. "The honest efforts of those involved in food production to correct bad practices are rarely noticed by the public," says Beryl, who worries that farmers are no longer heard. "Less than two percent of the American people are farmers," she tells me. "There are more environmentalists than there are farmers."

Beryl took me to meet Bill McCormick, the man in charge of the Buffalo Bill Dam. A no-nonsense engineer in a hardhat and cowboy boots, McCormick lives with his wife in a split-level house overlooking Cody, and in his off hours rides into Yellowstone on horseback. I trailed after him as he gave me a close-up tour of the dam, inside its trembling adits, around its toe and spillway, and up and down perilously small steel walkways to look at its vibrating tunnels and pipes. He told me not to fear the water that was absolutely everywhere, and showed me the stately brick-faced powerhouse that looked like a Carnegie Library cut into a cliff. McCormick has worked on the building of dams all over the West, from the Grand Coulee Dam to tunnels and reservoirs on the Frying Pan and Arkansas rivers. He helped train engineers brought over from Australia's Snowy Mountains Project. His years of working closely with the Shoshone Irrigation Districts make him uniquely qualified to discuss whether or not the farmers of the Shoshone Project ought to be on this land and whether the system is working well.

All in all he thinks it is. "The people depend on the systems," he mused. "They requested money to improve and maintain them. You know they would have given up if it weren't working. National politicians wouldn't have listened to three congressmen from the state of Wyoming requesting money." McCormick also knows that without the dam and irrigation canals, these farmers couldn't make it. "There wouldn't be much of our county. It would be mostly cattle country, but even then it would be limited because you couldn't raise enough hay to feed the cows."

Not all farmers are as successful as the Churchills and not all projects are as happy as the Shoshone. There are smaller irrigation districts with hard stories to tell, and there are some that haven't been managed well. "On the Shoshone Project, when the farmers took over the project, they became responsible for operation and maintenance," explains McCormick. "But many projects didn't maintain anything. And that's been the bad history of the irrigation projects. Some projects have been turned back to the government when operations went under. Others did a poor job of maintaining while keeping farmers' costs low. And now when they come to borrow that interest-free money and have to do environmental assessments, they scream bloody murder. There are districts that use too much water and there are some who whine. There are others who are grateful as hell. But not one district on the Shoshone Project has missed payments or is in arrears."[76]

The relatively small scale, one-crop-a-year farms of Wyoming are vastly different from the subsidized corporate spreads of California's Imperial Valley. Nothing makes Beryl see red faster than the mention of the word "subsidy." Her argument is simple. In 1979, the Garland Division entirely paid off their portion of the Buffalo Bill Dam and all of the costs associated with its diversion dams and tunnels, canals, and irrigation ditches. The project is prosperous, tax-paying, equipment-buying, and helps pay operation and maintenance costs for the dam and irrigation facilities. "We paid back every single penny," Beryl tells me. "We grow $42 million worth of crops on this land each year and we pay taxes on those crops. The only subsidy was that interest-free loan."

Bill McCormick doesn't shrink from the word "subsidy." "The big subsidy—a lack of interest on the loan—is a social issue to start with," he says. "The mission of the Reclamation Service was to make the West habitable by water development, and it will take a long time to understand if that was the right way to go. But Beryl and Winston are irrigation farmers. They were promised something that was correct at the time. They are good operators and they live well. But they pay through hard work and good management and investment.

"These are formidable obligations that the irrigation districts took on. They couldn't have done it by themselves. I'm the first to agree Western water users have been subsidized. But everybody in the country has been subsidized some way or another." McCormick feels that there are indirect benefits that most Americans reap from such subsidies—foodstuffs made available by irrigation, vegetables in January, grown in Texas or Arizona or California. "If you don't subsidize agriculture in some way," he says, "you're going to wind up with a few big companies similar to Dole in Hawaii or the big agro-business operations in California. If you eliminate little farmers and small outfits, you'll have more large financial institutions getting into it and we'll be at their mercy when it comes to prices. You'll also be at the mercy of natural disasters."

In front of the First National Bank in Powell stands a life-sized bronze statue of a man holding a shovel. Called the Desert Redeemer, the statue symbolizes these people and their feelings about their land. This is their place. They have worked tremendously hard to get it out of the grip of the desert and it has rewarded them. "Consider these figures," says Beryl

Churchill. "Total crop revenue through 1998 for the entire project was $720 million. When millions more are added for livestock revenues, the figures should reach $800 million. The Shoshone Area Project has paid for itself many times in the generation of business, purchasing power, and state, federal, and local taxes. The dollars loaned by the government should be considered the cost of the development of a nation."

Before I flew back to New York, I made a final pass by the Buffalo Bill Dam. The upper reaches of the Shoshone River still wash through the Absarokas and into a striking mountain lake behind the dam, only ninety years old but authentic-looking enough to now seem a piece of prehistory. Buffalo Bill Dam, hugging its mountain canyon, is higher now by 25 feet, in order to store a little more mountain runoff. "That world's fair feeling" is how Wallace Stegner identified an average visitor's reactions to Hoover Dam, but Buffalo Bill Dam is affecting in a more personal, handmade sort of way. It's not a big dam by anybody's standards these days, and that's at least partially why it's easy to love. But it's also a good dam, an enduring dam. It does its job.

All that snowmelt coming out of the Absarokas each spring, hoarded and sent rushing eastward through sluices and canals, is life itself to the hardworking farmers on the carefully tended benches of the Bighorn Basin. I left Cody and drove north toward Billings, Montana, watching the stark landscape roll past, mile after mile after mile. There was no water there, nor much human possibility. I remembered Bill McCormick's final words: "Reclamation developed the West. I think it's been good for the country. If it's the best financial deal for everybody, I don't know," he added tentatively. "I wouldn't touch that with a ten-foot pole. But without it there wouldn't be a Powell."

IN THE BIGHORN BASIN, in a cold corner of Wyoming, hardworking farmers carefully nurture one irrigated crop a year on semi-arid land. Halfway around the world, in Pakistan, irrigation canals slicing though even drier lands are farmed year-round. Farmers in each country confront many of the same problems. Pakistan's difficulties, however, are far more grave, since food for the entire country comes by virtue of irrigation in the scorched plains of the Indus Basin, a landscape altered irretrievably by engineers.[77]

Flying from Delhi to Islamabad, I caught my first glimpse of the dangerous dance between man and water in the Indus Plains. At first I couldn't understand what I was seeing. Splayed across the gray-green farmland below, a web of concrete arteries reached from horizon to horizon. Then I began to make out that the conduits, hundreds of feet wide, seemingly endless in length, were canals. Man-made rivers, powerful yet without a river's grace, the canals carry life itself across the flat reaches of a dry and gritty land.

After the partition of Pakistan from India, the Indus Water Treaty led to the construction of the world's largest contiguous irrigation system. Now, a century and a half after the British first began to build the canals, and only decades after they were completed under the Indus Treaty, Pakistan is in a state of near desperation over water. Every day I spent in Pakistan, my morning newspaper reported on a water problem.[78] Water disputes here are carried on at the highest level and at the lowest: between India and Pakistan; between regions, states, and provinces; between irrigation districts and cities; between villages; and even between farmers. Boys as young as nine have been jailed for water-incited murders. The importance of water, slow to dawn on much of the world, is already well understood in Pakistan.

In 1951, there were 34 million people in Pakistan. Fifty years later there are 152 million. Until now, the country has managed to sustain adequate agricultural growth to feed its exploding population.[79] Those remarkable increases in production came about through two singular and related efforts: the green revolution, with its artificial fertilizers and improved strains of rice, maize, corn and wheat, and the water brought by the massive physical manipulation of the Indus. In 25 years, at current birth rates, there will be 260 million people, but most of Pakistan's arable land and available water have now been exploited. Worst of all,

Pakistan, one of the most densely packed nations in the world, must now face the fact that the system it was given only 40 years ago is in calamitous straits.

Cotton, wheat, and corn grow wonderfully well in the warm climate and rich alluvial silt of the Plain provided there is water. But the only water here comes from the rivers of the Indus watershed, which flow full in the spring with runoff from the Himalayan glaciers. Before modern irrigation began in the basin, groundwater lay deep beneath the land. But canal loads of water from the Indus River have brought water tables nearly to its surface. Too much water, as damaging as too little, chokes off oxygen and life from the root systems of plants, interferes with the rotting of organic materials, reduces nitrogen, and encourages the accumulation of toxins in the ground. Almost a quarter of Pakistan's Canal Command area is waterlogged.

Rising water tables carry with them to the surface an unwelcome hanger-on. Even when water carries only a little salt, its repeated application in irrigation cycles dissolves more and more salts out of the soil or from saline springs or rock formations.[80] As the water at the surface of the land evaporates, an accumulation of salt is left behind, coating the earth in a deathly crust. Every year, 55 million tons of salt are carried onto the farmlands of the Punjab and the Sindh by canals, but only 11 to 16 million tons of that run off into the sea. Forty years after the Indus Water Treaty, a full quarter of Pakistan's crop potential is disabled by salt.

Waterlogging and salination are common to arid land irrigation but are especially tenacious in the Indus Basin. Hot weather, porous soils, and sluggish drainage on the only slightly inclined plain all favor rising water tables and salt accumulation.[81] As early as the 1960s, Pakistan found that it was losing a hundred thousand acres of farmland each year to waterlogging and salinity. The farmers of the Indus Basin have participated in a dismal saga of trouble. Water allocation is often faulty and occasionally nonexistent. Some land receives an abundance of water; other land gets no water at all. Farmers, who pay low, subsidized rates for water, often use it carelessly. Breached, broken, leaky, and clogged with silt, the whole system needs renovation. As little as 30 percent of all water diverted into the canals of the Indus Basin makes it to the root zones of crops.[82]

Since I've been told that waterlogging and drainage are technically solvable problems, I wanted to know if there was anything to be done about Pakistan's difficulties. No one in Pakistan denies their precarious position.[83] The Indus Basin is a virtual laboratory of water development. If international conferences, consultants, advisors, projects, and studies could staunch these wounds, Pakistan would be cozily well-watered by now. Obviously, even in a place as desperate for solutions as Pakistan, there are no easy answers. On a perfect spring day, I drove out of Lahore along a tree-lined canal, past apartment complexes, schools, boisterous children swimming in the canals, groves of flowering bushes, and apparently prosperous farmland. Pakistan, especially its Punjab, is physically enchanting—given the difficulties, almost painfully so. On my arrival at the headquarters of the International Waterlogging and Salinity Research Institute (IWASRI), I was met by several engineers who patiently explained the extent of their dilemma.

Their leader was the director general of the institute, F. A. Zuberi, a small, wiry, and bespectacled man who was quick to impress on me the urgency of the country's population problem. "In Pakistan, there will be ten thousand more people at the breakfast table tomorrow morning than today," he said frankly. "So water becomes scarce while sources of pollution increase. Requirements for industrial use and for municipal use also increase. All of these things taken together make a very sad picture." For more than 30 years, Pakistani engineers have been scrambling to offset the problems visited on them by the engineering works, while at the same time delivering more fresh water to their ever-expanding society.

The way to handle waterlogging and excess salt is to flush salt out of the soil and drain the land. Salted land can also be treated with chemicals to restore its structure, but it must lie fallow for some time. Drainage, lowering water tables, and letting the land rest seem like simple answers. But, of course, it's not at all simple.

"We know that in the future drainage works must precede irrigation works," William Willcocks warned in 1904, when the British were building these canals. Willcocks understood that a canal system should be shadowed by subsurface tile drains or pipes that discharge into a larger drain to take excess or salty water out of the fields. If the British or the Pakistanis had built a drainage system concurrently with the canal system and the canals had been lined, Pakistan would be a different place today. Now the costs of adding drainage are prohibitive. In a 1992 paper written for the World Bank, Masood Ahmad and Gary Kutcher reported: "This [drainage] amounts to a potential cost of about $9 billion for the approximately 8 million acres of canal command area affected— almost the total value of the agricultural sector's output in one year."

Even in places where there are drains, the water needs somewhere to go once it is removed from the fields. In the Sindh, at the receiving end of the canal system, brackish drainage water turns a bad situation into a lethal one.[84] Pakistan is at work on a 200-mile drain that will carry 11 to 16 million tons of salt out of the system every year into the Arabian Sea, and Zuberi tells me that they are considering extending this drain 600 or 700 miles farther to the upper plains.

Zuberi and his engineers are also building evaporation ponds—vast cement basins, some as large as 40,000 acres, for the disposal of salty water, through evaporation. Intended for uninhabited desert areas, the ponds take a long time from planning to implementation, and in the interim, many uninhabited areas have become inhabited. They hold a relatively small amount of drainage effluent; evaporation is reduced as the salinity becomes concentrated and seepage has been found around their edges. IWASRI has installed over 15,000 tube wells, which are used to lower water tables, but they often bring up water already containing salts, are expensive to run and maintain, and deteriorate quickly. However, if the pumping stops, water tables rise. When Zuberi's teams have applied money and effort, projects have controlled waterlogging and returned land to production. But in 30 years of rehabilitation they have overhauled only 8,300 miles of a system comprised of 36,000 miles of canal and a million miles of watercourses and ditches.[85]

Pakistan's engineers have prepared plans for more reservoirs to store water. "Two dams have been planned," Zuberi told me. "But this is for the politicians to decide." Big dams, at Kalabagh or elsewhere, will be built to meet the demands of Pakistan's dangerous population growth. The World Bank has already expressed its willingness to bankroll at least two. But dams are no answer to the shortsighted practice of pouring water into a faltering canal system without fixing it, or of bringing marginal land into production at the same time that good land is swamped by water or dying for the lack of it.[86]

POURING WATER on plants to keep them alive is as natural as milking a cow or sticking seeds into the ground. But anyone who has ever watered a house plant and watched white patches turn up on the sides of the clay pot understands something about irrigation and salt—that what is needed is enough water to keep the plant alive, yet not so much that it will turn into something resembling Lot's wife. Dry-land irrigation farmers from Asia to America do battle with salt, loads of it. Water dissolves salts, rivers carry them, and irrigation distributes them. One acre of agricultural land can disperse eight tons of salt into a watershed. The Food and Agriculture Organization of the United Nations estimates that worldwide 50 to 75 million acres of land are severely salted, and that another 150 to 200 million are partially affected. Some of the worst damage is in China, Pakistan, the Central Asian Republics, and the United States.

"We little realized how difficult it was to reclaim land quickly, and when reclaimed, to keep up its wealth and fertility," wrote William Willcocks. Salt damage should be stopped before it starts; by watering

sparingly, building drains, lining canals; leveling land so that water distribution is uniform; and maintaining distribution pipes or ditches.[87] "No
regime ever built a monument to itself with tile drains, but it is at that
level that Egyptian planners must focus their attention," writes John
Waterbury. Salt needs to be flushed out of fields in short, intense applications of water.[88] Planting salt-tolerant trees, such as *Prosopis juliflora,*
Eucalyptus, or *Acacia nilotica,* which sop up excess water and open up the
soil, can mitigate the level of salt. Wherever groundwater is plentiful, it
is important to pump.

None of these methods is without difficulty—canal linings break
down and drains clog—but they can be managed. When substantial
parts of the 500,000-acre Mahi Project in Gujarat, India, became waterlogged, its managers quickly reduced water deliveries and improved the
drainage system. Simultaneously, farmers began conjunctive use of tube
wells. Their combined efforts helped drop groundwater tables, and Mahi
Project crop yields are now among the best in the state. Monitoring,
control, and remedial engineering not only make projects work, but can
reverse damage. This is just one project, but our experiences with greening deserts around the world have yielded some common solutions.

Flood irrigation is still the usual way to water crops around the world,
but newer methods use far less water.[89] The Israelis, some of the best
technicians anywhere, have reduced the amount of water used on crops
by a third by using computer-controlled drip irrigation. In Kansas,
where 95 percent of crops are irrigated, farmers who had been using
center-pivot irrigation cut their water use in half by turning to subsurface drip irrigation and at the same time boosted their crop yields. At the
International Rice Research Institute in the Philippines, researchers
increased crop production by 39 percent with simple changes in water
distribution procedures and some technical improvements. The World-
watch Institute says that if we could improve worldwide irrigation efficiency by just 10 percent, it would save enough water to meet the
personal needs of every human being on the planet.

Small-scale irrigation projects are faster to build and implement, and they encourage closer social collaboration and local involvement.[90] Cost returns on a large project, however, are often two or three times that of small projects. Small irrigation versus big or government versus private—no single method promises success all the time in every place.[91,92] Each parcel of land embraces distinct and varied combinations of soil, slope, weather, and sources of water. The Churchills, in the snow-fed lands of Wyoming, face problems that differ widely from the problems of the distressed farmers of Pakistan, and their solutions are by necessity their own.

AFTER LEAVING ZUBERI and his earnest colleagues, I returned to Lahore to speak with a different kind of expert, a man as intimately engaged with canals as the engineers, but with a vastly different point of view. Syed Ayub Qutub was a boy when the Trimmu-Sidhnai Link Canal cut across his father's farm in the Punjab Province in the early sixties. A tall, fair-haired, Cambridge-educated man, the soft-spoken Qutub has been involved with water issues ever since the canal brought the first water and with it traces of Pakistan's agricultural decline. "After partition, we were not trusting of our neighbors," Ayub Qutub told me in the lavishly flowered garden behind my hotel. "Instead of a management solution—the sharing of our waters—we went into engineering solutions. People find it easier to work with materials than to work with people. And because there seems to be a kind of permanence in engineering solutions—especially between states—Pakistan thought it would only be secure if it had physically ensured that its canal commands fed from the eastern rivers were to get water from the western rivers. Therefore, we concluded the Indus Basin Works Program, which takes water out of western basins and cuts across the natural line of drainage to the eastern rivers."

Kaiser Engineers Pakistan built the Trimmu-Sidhnai Canal to carry irrigation water from a barrage on the Chenab River into the Ravi River.

To build the 50-mile canal, Kaiser excavated 20 million cubic yards of earth. "We owned a two-hundred-acre livestock farm next to the old Havali Canal, a lined British canal built in the 1940s," Ayub Qutub told me. "First our farm got lopped off by twenty acres and that was not so bad—but there isn't enough money to line these canals, so they are all, in Shakespearean terms, as leaky as an unstaunched wound. As these monsters leak, land on both sides becomes waterlogged. Then the water table rises." The land that Ayub Qutub's father worked and where he is now buried has become useless. Qutub, a man normally sheathed in a kind of Gandhian restraint, is driven to distraction by the chain of events that led to the ruin of his family farms. "If it is already marginal land, you must abandon it," he says angrily. "Which is what I have done. There are, along the length of these canals, thousands of farmers who were so affected—the undercompensated victims of the grand works."

Qutub's conclusions are startling, since he believes that the only way to save the largest irrigation system in the world may be working with its smallest, least significant components. First of all, he explained that while lining hundreds of thousands of miles of canals in the Indus Plains, even in brick, could cost as much as $74,000 a mile, it would bring almost 6 million acre-feet of water to the root zone of crops. That's nearly as much as a new dam at Kalabagh would hold in its reservoir.[93] Ayub Qutub and PIEDAR (Pakistani Institute for Environment Development Action Research), the nongovernmental organization of which he is a founder, have taken their faith in local solutions and have begun to line watercourses, the channels that deliver water directly onto fields. "Giant canals and dams are often built for reasons that a small investment in human trust and human community would have made unnecessary," says Qutub. "We can save all the water that we could store in a new big dam and we will do it at one sixth the cost."

PIEDAR, which organizes farmer communities in Pakistan's Punjab, gives loans to match village savings and offers technical training. "Surveying, leveling, and grading are critical skills in the Indus Plain, where slopes are small," Qutub explained. "You can take a man with ten grades of education, and in six weeks he can become a decent surveyor." He insists on the involvement of every member of a village, not just landowning farmers. "Farmers, whether they own land or farm land, do not live in isolation from the other elements of society," says Qutub. "The wandering cattle of the landless may break the bones of the watercourses. Women wash clothes on the banks of the watercourses. And funnily enough, even in this male-dominated society, women are the prime savers. You cannot mobilize multipurpose village organizations without mobilizing the women and bringing in the landless."

Qutub tells me that he and his coworkers—anthropologists, planners, and social organizers—make sure that the design of all water systems, including those of potable water, is done with and by the villagers themselves. "The government has delivered technically superb networks of water supply that never work, because no one owns them," he says. "But here, the village implements a water-supply scheme, which technically would make most engineers pull their hair out, yet is still the best. The villagers get in their hearts a warm feeling of participation, a sure feeling that what they are doing is accurate, and because the village itself purchases the bricks, cement, and sand, they keep an eagle eye on each other. Incrementally, you move to a perfect design rather than impose a design on a community."

I had learned from Zuberi that the government has also been involved in some on-farm management, and asked Qutub about their watercourse renovations. "God bless them, they are not a bad department," he says. "They have renovated 16,000 watercourses, out of around 90,000, in the last fifteen years. That is a thousand every year. At this pace, they will complete the course in ninety years and that is, of course, too late." PIEDAR began working with 25 villages in the Punjab ten years ago and has subsequently helped many more villages form farmer's organizations. He admits that this also is too slow a pace, but he fervently hopes that others will follow their more complete model and believes that what they have learned in the villages of the Punjab demonstrates a better way to work with an irrigation system.

Qutub told me that in 1992 he had headed a government-sponsored conservation study. What he learned in that study left him shaken. "It is possible for us to maintain our land and water resources and pull ourselves up by our bootstraps and become a Southeast Asia," says Qutub. "But it is equally possible, if we allow this to decay, to become another Somalia or Ethiopia. Both options are open and both are possible. But there is no staying where we are. Pakistan is on the knife's edge."

NO ONE COULD KNOW more about the Indus Basin System than the man who built it, Syed Kirmani, the chief engineer of the Indus Basin Plan, who travels to Pakistan regularly and is visibly pained by accounts of waterlogging, salt, and faulty water allocations. I went to talk to him at his World Bank office in Washington. "All of those problems are technically solvable," Kirmani told me. "It may require investments, it may require effort, but the problems can be addressed. There are many areas where they have been solved. For instance, in the Punjab at Sheikhupura and Gujuranwalla, the fields were once white from salt but you don't find that now. Pakistan could be one of the major producers of the world. The soils are fertile. The farmers are innovative. The potential is there."

Kirmani has been unstinting in telling the government of Pakistan what must be done to get the irrigation system operating properly. His specific suggestions range from more control works on canals to overhauling governing institutions. Most of all, he stresses that the Indus Basin is a single hydrological unit. "Every action that disturbs surface and groundwater resources in one part of the basin," he says, "has a positive or a negative impact on the other parts. It must be managed as one unit."

Kirmani's passionate belief is that Pakistan's most deadly trouble lies at the heart of the system, the government. "Yes, there will be a meeting, then a commission, then another commission, and a report will come, but there will be no implementation," he says sorrowfully. "Mismanagement is at the height of the problems and people accept it as normal. Now *that* is a tragedy."

JUST BELOW PAKISTAN'S southern border lies a drier and even more desolate landscape, the sand dunes of the Thar Desert in India. The Thar in antiquity was called Maroost'hali, compounded of the Sanskrit *mri*, "to die," and *st'hali*, "arid or dry land," therefore meaning "abode of death." Only the hardiest living things can survive here: warrior clans and nomadic tribes, gazelles, wild ass, and desert fox. The beautiful but forbidding red-brown landscape is broken by occasional scrub, seasonal watercourses, and villages the color of the desert light. Whole towns lie buried under the unforgiving sand, covered by dunes that have rolled over them like ocean waves. People grew crops here once: the remains of plowed fields from 2200 B.C. have been found by archeologists. But agriculture didn't last because there is virtually no rain nor any year-round river. Those few who live here manage by husbanding every drop—and by moving. Every winter, the camel trains move out of ancestral lands to find fresh grazing, the sound of the fading camel bells a signal that it is time for the desert to rest.

But the quiet of the desert has been recently ruptured by a system even less forgiving than the one Zuberi and his engineers are trying to keep together. Water has come to western Rajasthan through the broad channels of the Rajasthan Canal, an enormous engineering effort that shunts Himalayan mountain water hundreds of miles south from snow-fed rivers into burning sands.[94] The Indus Treaty allotted water to Rajasthan, and digging on the canal began in 1958. After forty years, work is still going on and the canal extends to the ancient city of Jaisalmer and even to Jodhpur at the desert's edge. Although earth-moving equipment sculpted the dunes into a canal shape, the real construction was done by 40,000 workers at a time, women carrying pans of cement on their heads and men laying brick with bare hands. The first water flowed in 1961, and by the time the system is completed—around

2006—there will be 5,000 miles of artificial channels carrying mountain water into the desert. It is the Thar's only perennial water source.[95]

The canal has begun to profoundly transform the desert. There is green in a place that hasn't known it for more than 6,000 years.[96] The first of two canal stages irrigates 1.3 million acres, bringing in a yearly crop worth hundreds of millions of dollars. New towns in the Thar are crowded with grain merchants, banks, and tractor agencies. Livestock numbers—mostly goats—have increased exponentially, and land prices have multiplied at least 25 times. Since the canal was built, the Thar has become the most populated desert in the world.

Indian government officials and publications express elation about the gardens in the desert, saying that the canal will one day be to Rajasthan what the Nile is to Egypt. But already, worrisome signs are emerging. I have been visiting Rajasthan for two decades and have long heard rumors about troubles along the waterways—about sand dunes routinely covering channels; about how, as canal channels progressed, they were broken behind and the water "stolen"; and about the corruption plaguing the waterways.

There was almost no protest in the sixties when the Rajasthan Canal was begun, probably because so few people would be displaced and environmental impact assessments were a thing of the future. But recently, the URMUL Trust, a non-governmental organization that grew out of a dairy co-operative, became concerned with farmers' complaints about the canal. Local farmers felt badly betrayed as they watched the best land being auctioned off to prosperous farmers from elsewhere.

Meanwhile, according to Sanjoy Ghose, a young, courageous URMUL leader, "Corrupt administrators had reallotted the same land to two, sometimes three people, leaving them with their life savings spent and caught in expensive legal wrangles."[97] Local officials took advantage of their power by offering to increase water allowances for "considerations"—bribes. Waterlogging has increased rapidly—at least twenty-five thousand acres were already permanently bogged by the early 1990s—and many families displaced and forced into poverty.[98,99]

The Rajasthan Canal has all the usual troubles to which a desert irrigation system is heir and more: defective distributaries, uneven water distribution, wasted water, poor drainage, dust storms, and an epidemic of water hyacinth—a thick, resilient weed now sucking at canals and clogging watercourses. In order to keep up with the cost of making their land workable, farmers are forced to grow cash crops, ground nuts or water-hungry cotton, which has become the largest crop grown here, requiring three or four times as much water as wheat. According to the government's own records, water levels around the canal have been rising an average of three feet a year, bringing with them salinity, mosquitoes, and malaria.[100] Some villages have already been abandoned at the same time new ones are being built.

When engineers began building the canal, the effects of irrigating a desert weren't fully understood. The Indian government saw the greening of the desert as an act of faith, a heroic effort. In some ways, the government has tried to work with the environment, attempted reclamation and preventive measures in the second stage of the canal by cutting water allocations, experimenting with different kinds of cultivation, and putting in subsurface and vertical drains. An eco–task force led by India's famously efficient military plants some 800,000 trees a year along the channels, using fast-growing, water-absorbent yet drought-tolerant eucalyptus for "bio-drainage" and salt-tolerant plants to stabilize dunes. These are admirable efforts, but they do not mitigate the harsh impact of the excessive numbers of people and livestock that are displacing the life of the Rajasthani Desert.

Balendu Singh Parmaar, a mustachioed thakur from Bikaner, shakes his head when I ask him about the canal. "The canal has brought prosperity to many and drinking water to people who need it," he says. "But it is a sadly mixed blessing. Every politician has changed the course of the canal. Corrupt construction bosses used inferior materials, so it was badly built, and salts are already destroying land, which cannot support so many crops. But it is also changing the culture of the desert. So many northerners have moved into the desert that the dress code and language of our people have changed in just a few years." Anyone who has seen Rajasthani women in their vivid reds and oranges, blues and yellows—stunning in the burned desert—or the brightly colored turbans and curled mustaches of the men, will understand that the passing of one of the most beautiful manifestations of a culture is alone cause for gloom.

"The whole ecosystem is changing," Balendu tells me. "Even the climate has changed. Sandstorms used to move and replenish the soil. But there are no more sandstorms. Since the sand in the atmosphere no longer filters the sun, the temperature has risen. Fine desert grasses like bekeria and sewen are being destroyed by the deep plows and will never return. Nor will the imperial sand grouse, which lives in them. There are now herds of blue bulls, which eat anything at all and which we never before saw deep in the desert, as well as scavenging wild boar. These animals come at the expense of the Indian bustard, chinkara, and blackbuck."

The blackbuck is among the loveliest animals anywhere, a small antelope with spiraling horns that leaps across the landscape so lightly it seems to fly. Chocolaty-brown with white bellies, blackbucks are sheltered in the Thar by a worthy people called the Bishnoi, a Hindu sect of desert protectors. Bishnoi women carry every fifth pot of water for blackbucks, keep aside grain for them, and protect the trees that shelter them. Even more exquisite is the chinkara, a tiny black-horned gazelle, tail-wagging and ear-twitching, a perpetual-motion machine. Both chinkara and blackbuck are easy targets for the traps and guns of the meat-eating, gun-loving farmers who have moved here. The animals of the desert are not the only creatures coming into collision with the newcomers pouring into this desert. The Bishnoi themselves and the herders who move each year with their camels and livestock are being squeezed out as once-empty lands fill up with fences.[101] "Every aspect of life is changing in the desert because of this canal," laments Balendu Singh.

Dry landscapes are also fragile, subject to drought, wind erosion, and extreme temperatures, both hot and cold. A desert is by definition a place where evaporation exceeds precipitation. As the government of

India channels water from the mountains into the heart of the desert, too much is lost to the sun in a place where irrigated agriculture is difficult to sustain. The same water might be put to better use in the cooler fields of Haryana and the Punjab and even the pipelines of Delhi, where politicians are already fighting each other over every surplus acre-foot.

Defying the limits defined by nature in the Thar or any other desert is no simple matter. Over 30 percent of the world's irrigated desert land is waterlogged. The districts in the Thar that are gardens at the moment will deteriorate quickly unless swift and serious measures are taken to alleviate the problems that could defeat the canal system altogether. This greened desert will not last even a short time, and many of its new inhabitants will be seeking new homes in a country in which homes are in short supply, as some have already been forced to do. And because the desert is so fragile, they will have devastated its natural life before they leave.

FARMING AND LIVING in dry lands over long periods of time can work. In the American Southwest, the Hohokam Indians farmed with irrigation for 600 years. Along the Min River Valley in China in 250 B.C., Governor Li Peng built a thousand kilometers of canals that are still in use. The Egyptians have used Nile water to irrigate through the ages of man.

Sustaining life under such conditions, however, requires careful balancing. Wyoming farmers on the Shoshone Project are limited by weather and water and do not push the land beyond its capacity. In hotter climates we often demand too much of dry lands. That demand is usually based on dams, water diversions, and irrigation projects.

In Pakistan, Ayub Qutub and F. A. Zuberi represent opposite sides of the debate over large waterworks. I asked both men if Pakistan's dams and canals should ever have been built. "Water is crucial for life in Pakistan," Zuberi answered. "If there was no irrigation system, Pakistan would have been a desert. The dams and canals—if we don't have them then we cannot feed everybody."

But Qutub disagrees. "The consequences of irrigation are well known," he said. "Ultimately, irrigation civilizations have destroyed themselves by waterlogging and salinity. These monsters are so visible. They provide so much employment, are so impressive, so majestic to look at. But their weaknesses are insidious, slow, and ultimately lethal. It's an arrogance of the ultimate order. Man's attempts to control nature are found increasingly unwise, increasingly unsustainable, never sophisticated or complex enough for nature."

When I asked Zuberi and his engineering colleagues about the terrible problems the canals have brought, they chorused, "For every problem there is a solution!" It is at once a noble and misguided claim. As Qutub reminded me, these engineers are good men, working hard to solve their inherited problems. They believe that with enough freedom and resources, they can solve the most terrifying situation. But their resources are finite. The government of Pakistan is hard-pressed, even with aid from donors, to line, renovate, drain, and keep free of silt over a million miles of canals, watercourses, and ditches. The system Pakistan was given could not have been workable over a long time span without unimaginable inputs of money.

The engineers are not wrong—they *can* come up with technical solutions. They already know how to drain and flush the soil so it is arable, to pump water to the Arabian Sea or to the top of Mount Karakoram, how to grow salt-resistant plants and build a huge reservoir that will store monsoon water and send electricity to Karachi. Give them enough money and support and they will do it. But governments in developing countries—with so many demands on what little money there is—have so far proved spectacularly unable to provide what is necessary to make large-scale irrigation work. It is comparatively easy to put up concrete buttresses and viaducts. But those cement structures disrupt natural systems with artificial conduits that need an amount of care, money, and good governmental management not usually figured into project plans. While there is never enough money, all too often there is corruption.

Above all, governments that build large irrigation systems have generally failed to see irrigation for what it really is—a social contract. They have overlooked the main ingredient in making any irrigation system function—the farmers on whom any system ultimately depends.

In the semi-arid landscape of Wyoming, farmers grow reasonable crops on carefully watered land. In arid Pakistan, the same thing can be done but at a high cost and only with great vigilance. Too much is being asked of a system that was not constructed to adequately address its environment's true complexities. And there are some places, such as the Thar Desert, that should not be irrigated at all.

"We must begin to question the entire engineering paradigm," says Ayub Qutub. "Question it in larger system terms, environmental terms, and at the bottom, question it in human terms."

Chapter 5

PREDICAMENTS

Power and Water

*The difficult problem is to determine what is the greatest good,
and whether the several goods are compatible or whether one will
destroy others.*

WALLACE STEGNER, *THE MYTH OF THE WESTERN DAM*, 1965

Standing in the offices of the Snowy Mountains Authority in Australia
a few years ago, I watched as suddenly all of the staff and visitors rushed
to the window. It had begun to rain. "Beautiful rain," I heard them sigh,
almost as a group, "beautiful rain." For as many years as I'd been thinking
about water, I realized that in my own green landscapes I had rarely run
to a window to look at rain. In Australia—in the words of one biologist,
"by far the smallest, the flattest, the driest, the least fertile and climati-
cally the most unpredictable continent"—every downpour is an event.

One hundred and fifty miles inland from Australia's eastern edge, a
great ridge of mountains mimics the shape of the coast, from Victoria in
the south to Cape York in the north. This is the Great Dividing Range,
and its ridges catch airborne water blowing east from the Indian Ocean
then drop it over the glacier-sculpted snowfields of the Snowy Mountains. Australia's Snowies harbor one of the lustiest engineering schemes
of the last century—a supple engineering manipulation of geography,
gravity, and natural resources. Aqueducts climb over mountaintops;

power plants and tunnels are secreted deep beneath solid granite, and artificial reservoirs cradle the mountains' watery wealth. I'd heard about the daunting work of Sir William Hudson, builder of the Snowy Mountains Hydro-electric Scheme, and decided to spend some time in Cooma, Hudson's headquarters at the base of the mountains, to explore the outcome of his project.

These mountains make an American think first of Montana or Wyoming—there are wild horses, drovers, green hills, and snowcaps. But instead of pine, sagebrush, buffalo, and bears, you will find gum trees, wattle, kangaroos, and wombats. Several large rivers wind unpredictably through the forests and alpine meadows that lie in the shadow of Mount Kosciuszko, Australia's highest peak at over 7,500 feet. None twist and turn more unpredictably than the Snowy River, fed by a bewildering number of tributaries, undulating its way north, then south, and back again, before finally plunging south into the Tasman Sea. The Snowy River's furious torrents once carried more water through sloping gorges than any other Australian river.[102] That was before the Snowy Mountains Scheme caught the river at Island Bend and sent it due west.

THE TOUGH FARMERS in Australia's parched eastern coastal provinces were the first to turn their eyes covetously upward, toward the rain-fed Snowy Mountains, where an abundance of water dropped indifferently into the sea. Persistent, searing droughts had tormented and shaped the mind-set of Australia's European settlers from their earliest days on this continent, its original inhabitants—aboriginal hunter-gatherers—staying within walking distance of rivers or ponds. But even though the livestock-heavy Europeans pondered the idea of mountain reservoirs to their fields, it was not their desire for water that incited the building of those dams. It was electricity that finally prompted the building of the Snowy Scheme.

I hadn't been in the Snowy Mountains very long when I learned that Australians had shared with Americans a kind of frontier daring in the nineteenth century, and that they had been well ahead of the field in taking electricity from water.[103] In 1883, in Queensland, a water-driven wheel generated electricity for the town of Thargomindah, only a year after the first turbine generating plant using impounded water came into operation in the United States at Appleton, Wisconsin. Around that time, one optimistic surveyor proposed that the Snowy River's "unutilized dynamical energy" supply the entire power needs of the Commonwealth. In 1947, after a century of expeditions, surveys, conferences, public meetings, and disagreements over the Snowy's rich reserves of water, engineers eagerly determined that it would be possible to turn the Snowy River around, tunnel it deep beneath mountains, then drop it again and again, generating hydropower at every fall until it reached the Murray, the Tumut, and the Murrumbidgee rivers, where it would irrigate crops. Canberra and Sydney, which then suffered frequent blackouts, would receive the Snowy's electricity, 3 million kilowatt hours, the power produced by 4 million tons of good coal a year or 0.5 million gallons of crude oil per day.[104]

The conflict over the development of the Snowy River was as fierce as the battles over America's Tennessee Valley Authority. Everybody wanted a piece of the Snowy, from states to town councils. The pertinent contenders were New South Wales, where the Snowy begins, which wanted Snowy waters diverted northwest to the Murrumbidgee River mainly for irrigation, and Victoria, where the river exited the continent, which wanted the river sent southwest to the Murray River, mainly for electricity. To get the $500 million project through Parliament, the Snowy Committee Chairman, Nelson Lemmon, argued that power was needed to defend the country. Japanese submarines had actually entered Sydney Harbor during World War II, and Lemmon pointed out that Australia's coal-fired power stations, located along the coast, made easy targets. "They could blow all your damned electricity out in one night's shooting!" he thundered to Prime Minister Ben Chifley. "Where'll you produce the arms? Where'll your production be with all the power of New South Wales buggered?"

Lemmon won that argument and Chifley agreed with him. But who would build it? Who would pay? Who would control its benefits? After more haggling, a refined plan was brought before Parliament and some well-organized opposition. A New South Wales minister who felt the Commonwealth was interfering with states' rights told Nelson Lemmon, "I'm sick of coming to Canberra to listen to your bloody pipe dreams. This scheme won't work and we're going to stop it any way we can."

"The Snowy Mountains plan is the greatest single project in our history," countered Prime Minister Chifley. "This is a plan for the nation, and it needs the nation to back it." Ultimately, the nation backed its prime minister. The Snowy Mountains Hydro-electric Power Act passed Parliament in July 1949. The Snowy Act established an independent Hydro-electric Authority, closely akin to the TVA, to be responsible for the design, construction, maintenance, operation, and protection of storage and hydroelectric works in the Snowy Mountains. Electricity would pay for the scheme. Irrigation was a bonus.

It was one thing to authorize one of the most daring engineering projects ever devised, another to actually get the job done. Across 2,500 forbidding square miles of the Snowy Range, heavily forested mountains dropped quickly into treacherous ravines and gorges. Swamps were infested with leeches, mosquitoes, and snakes. Tunnels would have to be blasted through solid granite, and caverns for power stations excavated out of that same hard rock. The climate was hellishly hot in summer and arctic-cold in protracted winters, and the builders would face downpours, ice storms, lightning, howling winds, snow, and mud.

Chifley worried about finding a man tough enough to see things through. Then William Hudson turned up. He was an unassuming fifty-three-year-old engineer with a long jaw and a boyish hank of hair across his forehead. Hudson had scribbled his qualifications on the back of an envelope. His coat was wrinkled and he wouldn't answer Chifley's questions. "I don't like talking about myself," he said flatly. Nonetheless, this shrewd, well-educated New Zealander had worked on dam and hydro projects in England, Scotland, and Australia. The prime minister somehow immediately understood that he had found the man for the job and offered him a salary bigger than his own. The Snowy Mountains Scheme was in the hands of a relatively unknown engineer who admitted that it wasn't going to be easy. "Ahead of us lie many years of toil, numerous obstacles to be surmounted, and, I have no doubt, many disappointments," William Hudson announced in a radio broadcast. "But the nation has accepted the Scheme, and if I judge Australians rightly, we will see that it goes through."

Hudson had, if anything, understated the challenge, as I learned from one of the original engineers, Alan Frost. "It was wild country," said Frost, a courtly, gray-haired Englishman who had shipped out to Sydney in the project's early days. "There were no—I repeat—no roads into it. The only people who had even a rough idea of the mountains were a handful of copper miners and graziers who had herds up there. You couldn't get a map because there weren't any. One of the first things Hudson did was organize an aerial survey. By 1950, he had produced a full set of contour maps so that at least we had the basics."

Hudson moved his headquarters to Cooma, at that time an isolated town whose only source of electricity was a handful of diesel engines. He started building barracks in the mountains for the army of hydrologists, geologists and surveyors who would make their way through the hills on skis and horseback. "They had bits and pieces of information and took potshots at levels which were fairly accurate, but there were very, very few trigonometry points," Frost recalled. Frost, whose engineering experience included projects in Scotland and India's Northwest Frontier, was made engineer-in-charge of the Electrical Division. "How the water came into the rivers, nobody knew," he remembered. "In 1950, Hudson sent a hydrological team into the mountains. People lived in tents and took constant readings to get the basic data on water flow. Before the year was out, they had 250 measuring stations on every major tributary."

Detailed surveying was critical. Since tunnels would be dug from at least two faces, both horizontal and vertical tunnel measurements had to be exact or the tunnelers might bypass each other in the rocky depths. There weren't many people living in the mountains—a few towns with a handful of stockmen or copper miners and their families—so there wasn't much relocation.[105] But hundreds of miles of roads would be needed for work camps with distinctively Australian names—Junction Shaft, Bugtown, Mudholes, and Gang-Gang. By extending existing skiing trails, the Snowy Mountains Authority could get to the headwaters of the Snowy River with relative ease. Frost's first project was to build a small concrete gravity dam on a tributary, the Guthega River, with a 3-mile tunnel, a 60-megawatt power station, 17.5 miles of aqueducts, and 50 miles of transmission lines, which would run straight to Cooma and light up Hudson's offices.[106]

But for all the geographical and climatic challenges Hudson faced, his greatest obstacle was a shortage of skilled help. Australia simply didn't have enough engineers. "We didn't have people available to do the detailing design," Frost recalled. Hudson advertised around the world and then approached the world's most experienced dam builders, the Bureau of Reclamation in the United States. The Bureau would send dozens of engineers to Australia to design dams and power plants, and more than a hundred Australian engineers came to Denver, Colorado, for training.[107]

Doug Price, a tall, handsome Australian, who joined the Snowy early in 1950 to help with preliminaries like stream-flow assessment, road building, and diamond drilling, recalls being sent off to get some hands-on experience building American dams. "There was a great affinity for Australia left over from wartime," recalls Price, who now runs a consulting firm in Cooma. "Working with the Bureau of Reclamation was great, and the Americans thought so too. We fit like a glove."

Hudson's next challenge was to find workers. There were only eight million people living in Australia then. Still recovering from the war, the nation felt vulnerable, underdeveloped, and short of muscle. Yet it was about to take on a construction project that would have given pause to any country in the world. The Snowy Scheme, born out of Australia's postwar edginess, found the solutions to its completion in Europe's postwar dislocations. The Old World was awash in refugees whose homes or families had been destroyed, who wanted to begin new lives, or were running from inhospitable regimes or wrecked economies. When the UN Relief and Rehabilitation Administration asked Prime Minister Chifley if Australia would accept a hundred thousand of them, he accepted immediately. Into Cooma and the mountains—populated almost entirely by descendants of the British and a small aboriginal population—flooded Italians, French, Austrians, Spaniards, Greeks, Egyptians, Russians, Ukrainians, Lebanese, Hungarians, Finns, and Jews. There were even former SS troopers, identifiable by blood-type tattoos under their arms. Men who hated each other, had perhaps shot at each other, now sat across from each other in dining halls and bars and rode together on small lifts into deep mountain pits.[108]

At first the men bunked together according to nationality, but after a large group of Poles had to be rounded up for hurling bricks into German barracks, Hudson decided to mix it up. "You won't be Balts or Slavs," he told his workers. "You will be men of the Snowy." Remarkably, they became just that. In *The Snowy: The Making of Modern Australia,* Brad Collis tells of two men who, while watching a film in a mess hall about the evacuation of Dunkirk, discovered that one of them, a German, had bombed the very jetty that the Englishman beside him used to escape to his boat. "The two men were seen some hours later leaving the canteen," he writes, "an arm over each other's shoulder."

Phone boxes in mountain towns carried instructions in a dozen languages. The children of immigrants found themselves playing with the sons and daughters of graziers and miners in mountain schoolyards. Except for the Serbs and Croats, who never managed to make a truce, these different people largely got along. It was their chance to make a new life. These men who proudly called themselves the "New Australians," made up two-thirds of the workforce that turned the Snowy Mountains Scheme into a reality. "I have been in engineering and construction since 1923 and I have never seen such a good team of men," said William Hudson.

Hudson himself was an extraordinary team leader. Doug Price, who

worked with Hudson from the beginning of the project, remembers that
he was "an inspiring chief, hard driver, absolutely fair, the kind of fellow
that could talk to the top and talk to the bottom in the same way. Hudson was the heart, soul and inspiration of the works, driving himself and
the Scheme toward the goal he saw so clearly, and expecting, even
demanding, that everyone else associated with it did likewise."[109] Price
says that one of Hudson's most innovative actions was to establish engineering laboratories in Cooma—"to keep the Authority up to date with
worldwide practice in fluid mechanics, physical sciences, geology, soil
conservation, and in materials investigation, but also to enable the
Authority to initiate new design and construction techniques."

"The Snowy Mountains Scheme is as bold in its thinking and as revolutionary in its consequences as the Tennessee Valley Authority," wrote
Barbara Ward in the *New York Times* in 1951. "What is probably the
largest single development project under construction in the free world
has started on its way in Australia with remarkably little public recognition of its significance."

In 1953, Hudson put out bids on four jobs designed by the Bureau of
Reclamation on the Upper Tumut River, north of the Snowy. The Upper
Tumut Project would create a gigantic reservoir behind the Eucumbene
Dam, the central storage for the Snowy Scheme, a reservoir nine times as
big as Sydney Harbor. To carry water from the Eucumbene Reservoir to
the Tumut River, a fourteen-mile-long tunnel was to be blasted 1,800 feet
underneath the mountains, to drop Lake Eucumbene water 2,700 feet,
down through three power stations.[110]

An American consortium, Kaiser, Walsh, Perini & Raymond, led by
Henry Kaiser's men, won the bids on Eucumbene's hydraulic tunnel
project and the building of a sloping concrete arch dam at Tumut Pond.
The "Yanks" immediately set up a railhead at Cooma and hired tunnelers,
or "molemen." Special jumbos—the multitiered drilling rigs used at
Hoover—were built to fit the twenty-two-foot-wide tunnel. Kaiser
started boring in November 1954, putting out the word that they needed
to get underground before winter snowstorms hit.[111] By April, just as it

was starting to get chilly, they were already seventeen hundred feet in. Ten miles away, a shaft had been dropped 300 feet to the main tunnel to lower men, equipment, and an entire locomotive with tracks, so that blasting crews and molemen could start working back toward the tunnelers coming in from the entrance.

The Americans brought with them rigorous work requirements. The Snowy's contracts carried a $2,800 penalty for each day the job went over schedule but a bonus of $1,400 a day for bringing it in early. The Yanks wanted the bonus, and they worked around the clock in three shifts, six days a week, launched at the firing of a gun. "The Americans showed us the essence of driving a contract through," said Snowy Mountain Authority Safety Inspector Colin Purcill. "They also showed us competitiveness. Not a lot of us liked their bull-gang approach—this bellowin' screamin' caper—but certainly things started movin'."

"They paid well, accommodated well, and they expected their people to work," says Doug Price. "They gave people rough-weather gear, and if it was raining they expected them to work in the rain. Some of it was not quite what had been done in Australia up until that time. But I think it was a good deal for both sides and it was for the ultimate good of the country."

Wages were three times that of Australian underground work. Tunnelers were given a mileage bonus, and the teams of molemen broke international tunneling records again and again. The original contract stipulated weekly progress of a mere 69 feet, but Kaiser's supervisors worked the men up to 484 feet a week. When a record was broken, Kaiser put on special meals for everyone, with plenty of beer.

Kaiser's people found the dam building fairly easy going, but blasting the underground tunnels was treacherous. Drillers attacked the rock face and then loaded the holes they'd drilled with gelignite, often by locomotive light, since electric lights could spark explosives. Untrained workers were watched closely. A tunneler named Ken White described what it was like being a new man underground to the writer Siobhan McHugh: "It was quite an unusual feeling, going down into the unknown. It was black, and nobody spoke, nobody moved. They all just looked very blank, very solemn, and they had these terrible yellow clothes—wet-weather gear actually—that was all covered in mud. We got in these little carriages and we rattled our way down this tunnel, which was very small, and we thought, 'My God, is this the size of this?' because it was barely missing our heads. . . . Then, after a while, they were blasting further down. . . . Somebody sort of grabbed hold of me by the shoulder and pulled me down to the ground and made me hide behind a railway truck, and I didn't know what was going on at all, because everybody put their hands over their ears. Then there was this mighty loud bang, and this wind came up the tunnel so fast . . . It was obviously a shock wave from the blast, so as soon as that happened, I thought, 'Oh well, that's all over,' and I stood up—and then the wind came back again and blew me over!"[112]

One hundred and twenty-one men died in the twenty-five years it took to complete the project.[113] The Snowy deaths were grisly. Fifty-three died underground, caught by ill-timed dynamite blasts, leveled by falling rocks, or crushed by moving trains. Other men died falling down shafts or toppling from scaffolding. They were hit by steel pipes and pieces of concrete, electrocuted, burned, drowned, smashed by earth-movers, or hit by trucks. Many more were disabled.

Work on the surface was dangerous too, even on a good day. Tractors turned over, things fell. There were gale-force winds. When blizzards buried work camps, help was sometimes days in coming. Winter light-ning season was especially frightening since electrical storms could deto-nate explosive charges prematurely. For a long time there was only one physician on the Scheme, a Romanian named Dr. Ina Berents, but a vet-erinarian in Island Bend took on some of the lighter human work.[114] Whenever a man was killed, everybody went into Cooma and drank all night. It could so easily have been any of them. One Russian worker said, "The only time we ever got to know one another was when there was a death. It was like war."

The first victory of the war in the Snowies came in the form of electricity, lots of it. By late 1954, Guthega was about to come online. "After three years of construction, we were ready to go into business," Alan Frost told me. "But there was an almighty drought. The dam was complete, but it took us six weeks to get enough water into the reservoir to flood the pressure tunnel and start the power station. We commissioned the power plant before Christmas, and in January sent the output through to Sydney. That winter, that one power station of ours saved Sydney from blackouts forty-seven times." In 1955, Queen Elizabeth II made William Hudson a knight of the British Empire.

Meanwhile, Kaiser, Walsh, Perini & Raymond was at work on the Tumut Pond Dam. Doug Price had been given the job of supervising it. "All these fellows had been together on Bonneville or Hoover or Coulee and spoke the same language," Price recalls. "We'd be on some problem at Tumut and the guys would say 'Well, when we was at Boulder . . . ' And I'd say 'Well, you're not at Boulder, and here you're gonna do this.' "

At the same time, after six years at the largest dam, Eucumbene, the New South Wales Works Department hadn't managed to get much done, complaining that it couldn't control the workers. Hudson fired the department heads and put out public bids, which were won by Kaiser. Kaiser immediately fired all the laborers and hired new ones. Eucumbene's foundations were in place, but Kaiser now had to quickly raise the world's second largest earth and rockfill dam.[115] This meant blasting and transporting 7 million cubic yards of earth and rock to the dam site— the rough equivalent of picking up a small mountain and moving it down the road. Kaiser's people used coyote blasting, which involved burrowing strategically located small tunnels into a mountainside and loading them with explosives. The preparation of one blast took at least a month. The first explosion used 19 tons of dynamite and liquefied the better part of a good-sized hill. Ten such blasts produced 2.6 million cubic yards of quartzite rockfill. Debris-filled trucks made thirty trips a day between blasting sites and the dam until Kaiser improved roads so that they could make sixty. In 1958, Kaiser, Walsh, Perini & Raymond completed the Eucumbene Dam two years ahead of schedule. They also finished the 14-mile-long Eucumbene-Tumut tunnel 4 months ahead of schedule and brought in Doug Price's finely curved Tumut Pond Dam.

At the dedication of Tumut Pond Dam, Prime Minister Robert Menzies, who had initially opposed the project, admitted that he had been wrong. "In a period in which we in Australia are still, I think, handicapped by parochialism, by a slight distrust of big ideas or of big enterprise, this Scheme is teaching us and everybody in Australia to think in a big way, to be thankful for big things, to be proud of big enterprises, and to be thankful for big men," he said.

It took big, tough men to build this system. When the second group of major contracts went out for bids in 1958, Australian companies, having gained skills and engineering sophistication, won most of the jobs.[116] The homegrown companies broke tunneling and earth-moving records, developed new diamond-drilling techniques and high-speed turbines and transformers. They introduced 330-volt electrical transmission lines at a time when Europe was still using 220. The Authority also pioneered new rockbolting techniques, a method of securing unstable rock in tunnels by drilling a horseshoe pattern of closely spaced holes, inserting steel rods, and then double-grouting them. The technique is used worldwide.[117]

As Australians played a larger and larger role in the project, the Snowy captured the national imagination. "Snowy River, Roll," a song written by Bill Lovelock, a Snowy worker, became a nationwide hit:

> Give me a man, who's a man among men,
> Who'll stow his white collar and put down his pen.
> We'll blow down a mountain, and build us a dam,
> Bigger and better than Old Uncle Sam![118]

"The relationship with the Bureau of Reclamation went on for a long time," says Doug Price with apparent pride. "But by the late sixties, I suppose we had grown up a lot ourselves and had developed our own expertise. In the process, the U.S. Bureau was getting some information from us." Australian companies built most of the Snowy.[119]

Between 1965 and 1968, drought gripped southeast Australia, but even though the project was unfinished, in the irrigation areas served by it, crops and livestock were spared because of blessed Snowy water. The project had begun to work for the country from the time its first dam was built. The Snowy Scheme was completed in 25 years. Its 16 big dams, backing up as many massive blue reservoirs, sprawl across the mountains linked by 90 miles of tunnels, 7 power stations, and 50 miles of aqueducts and pipeline. "After the Snowy," said Nelson Lemmon, "she was a nation."[120] The Snowy Mountains Scheme was born out of wartime insecurity, fear of invasion, fear of drought, and, above all, a need for electricity. It became a matter of national pride and by the time it was finished, it had helped to develop eastern Australia. Because it was working so well, people eventually forgot it was there.

FOSSIL FUEL PLANTS are impossible to forget. The burning of coal and other fossil fuels like gas and oil sends over 800 tons of carbon dioxide into the air each second. That's 27 billion tons of carbon dioxide a year. The burning of one ton of a fossil fuel results in the buildup of several tons of carbon dioxide, and it is this buildup in the atmosphere that is speeding the warming of our world.

Every energy source comes with some sort of unpleasant price. Nuclear energy is risky, thoroughly uneconomic, and there is no satisfactory way of disposing of the waste. My family home is near Spring Valley in western New York State, where authorities built a "state-of-the-art" nuclear-waste storage site. A few years after storage began, somebody discovered radioactive hair clippings on the floor of the local barbershop.[121] Natural gas is cleaner than coal by almost half, but it adds hefty amounts of carbon dioxide to the air, and gas plants, now going up in large numbers, will warm the atmosphere as well.[122] Coal, even what industry people call "clean coal," produces the most damaging and noxious emissions and the most carbon dioxide.

Our desire for electricity is growing fast. Until I began work on this book, I never really noticed electricity. I presumed that it would always be available in the same way that I expected water and air. That began to change about a decade ago, on a winter afternoon in central Turkey, as the town's daily allocation of power ended. Life came to a virtual standstill in the cold and dark, as bodies faded into the shadows and the only visible brightness came from the headlights of a few scattered automobiles. Not long after that, driving at night in Assam through a blacker darkness than I'd ever seen, I was confounded by the tiny pricks of light from the single kerosene lamp in every village home. There are villages just like these throughout the developing world—40 percent of the world is without electricity—and the lives of the people in most of them would be better for having it.

Right now, the generating capacity of the entire underdeveloped world—80 percent of the world's population—is only two-thirds that of the United States. Half of that population is in China and South Asia, both of which use coal for most of their energy. Emissions from coal-burning power plants being built to serve those two regions alone will soon double, casting a dark, gritty shadow over the whole world in the twenty-first century. Energy use around the world has increased by 70 percent in the last thirty years, and if demand grows as it is expected to, greenhouse-gas emissions will increase by 50 percent by the end of the next decade. Even if we could hold our emission levels to their current output, there will be twice as much carbon dioxide in our atmosphere at the end of the twenty-first century as there is at its beginning.

We badly need an energy source that doesn't heat up the atmosphere, or shorten life expectancy from its toxicity, and we cannot afford to overlook any source of clean, renewable electricity.[123] But none of the renewables have yet been perfected. Solar and wind energy systems, because they are not yet mass-produced, are still expensive and take up a lot of ground. Power from hydrogen-powered fuel cells is excessively costly because it requires energy to split water molecules. Thermal power requires water to recharge steam. Wave machines cause havoc in underwater environments. Biofuels require a great deal of land for production and release carbon dioxide into the atmosphere.

Whoever resolves the problems of clean energy will have a tremendous competitive edge in the future. Already, big energy producers like Shell, Canon, and BP Amoco have entered the field. British Petroleum, whose management has formally recognized climate change, says that its investments in solar technologies alone will come to $1 billion in the next decade. Royal Dutch/Shell predicts annual sales in renewables as high as $50 billion in the next twenty years. Shell, DaimlerChrysler, Ballard Power systems, and Norsk Hydro have combined forces with Iceland in a countrywide effort to create a nonpolluting, hydrogen-powered economy with the intention of eventually phasing out fossil fuels altogether.

In the last ten years, the same decade in which carbon dioxide emissions in the United States increased by 15.5 percent, the costs of practical, renewable energy sources have dropped like hailstones. Solar, wind, hydrogen, fuel cells, bio-gas, and geothermal energy are gradually becoming viable. (Some Midwestern farmers have discovered, for example, that it is more profitable to sell electricity from wind farms than to grow crops.)[124] But it's sadly true that at this moment, renewables are not yet making a significant contribution to the enormous and growing worldwide demand for power. We need some time. The single renewable source of energy that can deliver the large amounts needed is hydropower, and the Snowy Mountains Scheme demonstrates it at its best.

THERE WAS A STORM in the mountains on my first day's drive through the Snowy Mountains, bringing trees down on the highway that snaked treacherously through clouds and fog. Snowcaps towered over me—the natural water storage fields of the Snowy Mountains. I was driving across the mountains to explore the end product of the Snowy Scheme and was surprised to find that except for the clear mountain lakes, it was nearly invisible. Only two percent of the mountain landscape is burdened with engineering, and the little that isn't underground is masked by eucalyptus trees.

One of the Scheme's most unexpected outcomes has been the preservation of the mountains. The Snowy Mountains Authority planted trees vigorously and put clauses in all major contracts that builders were responsible for reclamation projects after construction. The most critical element, however, was the removal of cattle. George Seddon, an Australian environmental writer, notes that the original cattlemen from the Snowy River area, despite being resourceful horsemen, "were mostly illiterate, exploitative, and very destructive." Seddon says that the cattlemen rendered large tracts of land "useless within a few decades." But because the Snowy Mountains Authority was ahead of its time in environmental matters and concerned about sediment in the reservoirs, Hudson worked hard to get grazing leases canceled, a controversial move but important because, unlike the padded feet of Australia's indigenous animals, the hooves of cows and horses are devastating to soils weathered thin in a severe climate. Grazing and annual burn-off to clear the way for new grass for livestock was and remains the greatest cause of soil erosion in Australia.

Later, I flew back and forth across the mountains in a helicopter. From above, I could see ugly tunnel spoilage in a few places, but by and large the Scheme has folded into the forests. Most of the reservoirs lie within the six thousand square kilometers of the public lands of Mt. Kosciuszko National Park, and the Authority and the Park work together to keep the landscape pristine. The non-native, ornamental plants that Snowy construction workers brought for their gardens are a lingering annoyance. It may seem odd to get worked up about some wayward daffodils, but park rangers earnestly restrict the area to native species, which also means eliminating feral pigs, cats, and "brumbies"—wild horses that escaped from cattlemen's fences generations ago. We spotted small herds of the lovely brumbies from the helicopter, some with foals. Roger, the pilot, who used to muster horses with a helicopter in western Australia, told me that the farmers in the nearby plains sometimes let out stallions to fortify the wild populations. The Snowy Authority's most treacherous impact up here has been the roads, which opened the area to intruders—skiers, campers, fishermen, and hikers. Yet I drove long distances without seeing another car, and the tourist towns of Jindabyne and Thredbo, squatting pleasantly in the hills, seemed relatively benign.

The physical plant of the Snowy Scheme remains in good shape. At least twice a week an engineer walks across each dam and sends in a report. There are yearly inspections by electrical, mechanical, and civil engineers, who take readings, check pore pressure for water saturation and formations for erosion. Tunnels are drained and cleaned, and every five years there is an external audit.[125] The Snowy reservoirs don't contain much silt, partially because the granite beneath them doesn't break down, but also because the low levels of human intrusion mean that there is little erosion. Nor are there problems with eutrophication, since the reservoirs aren't highly vegetative. At least half of the reservoir water turns over each year, so there is no stagnation in the storages.[126] These dams are here for the long run.

"I think five hundred years is a reasonable life span," said Jack Grimstead, a craggy Australian of Norwegian descent, who is in charge of the physical works of the Snowy Authority. "There is a long horizon. But the unique part about the Snowy is the way the parts work closely together." The Snowy Scheme is a refined piece of engineering—handsome, tough, and flexible. Its giant valves and computerized power stations open, water pours, and turbines roar. Gates close to contain water in whichever reservoir the water liaison officers determine to be appropriate for the good of householders in Canberra, businesses in Sydney, or farmers in the plains.

The flexibility of the Snowy Scheme lies in its connecting tunnels and pressure pipelines. The Snowy reservoirs, scattered across thousands of miles, are a series of interconnected bathtubs, filled in the winter and spring with snow melt, then systematically emptied to generate power and water crops in the river basins of the Murray, Darling, and Murrumbidgee. Because most Snowy water is stored in one area—the Eucumbene Reservoir—a complicated water accounting system has been developed in which hydrographers measure inflows, water balances and information on current diversions, and bring the figures back to Cooma to use in their monthly accounting. Water liaison officers develop an annual plan that can weave its way through extremely wet or dry conditions and still meet electricity and irrigation requirements.

The Scheme's power plants, situated midway between Sydney, Melbourne, and Adelaide, each year generate roughly 5,000 gigawatt hours of peak power—electricity needed at periods of peak demand. Its 7 power stations produce 16 percent of the capacity of the southeastern Australian grid. This may not sound like much but the Snowy power plants are important for flexibility since they provide a fast response. Hydro plants can be up and running in minutes to pick up failures or demands in the grid. They also change output rapidly—either adding or dropping power—so the total generation on the system always matches total load. And since a hydro plant can start up without external electricity, if there is a system failure the Snowy can jump-start the whole grid.[127] Coal-fired plants take hours to come online.

Standing on the floor of Snowy's Tumut 3 Power Station, I felt it quiver under me as great quantities of water from Lake Talbingo moved in and, within a minute and a half, began turning turbines. The Snowy's fast response is still called upon regularly—in 1996 alone, it responded to 44 grid emergencies. More significantly, coal-generated electricity comparable to the Snowy's output would yearly burn more than 2 million tons of black coal, or 8 million tons of poorer quality brown coal, and dump 5.5 million tons of carbon dioxide into the atmosphere along with 16,000 tons of sulfur and 11,500 tons of nitrogen-oxide.[128]

Worldwide, the failure of hydropower has been its inability to compete in power markets. When the Snowy was built, hydropower was cheaper than thermal power in Australia, but gradually the cost of coal-fired electricity became lower. But since the damaging effects of emissions from coal-fired power plants are consistently left out of the cost equation, hydro plants have generally not been justly evaluated against the less attractive aspects of their competitors.

Electricity generated by the Snowy Scheme pays its operation costs and repays the construction loans. The Snowy also makes a significant contribution to upward of $7 billion worth of crops grown each year in the Murray-Darling Basin. "In average years we don't contribute much at all," soft-spoken Water Resources Manager Barry Dunn told me in Cooma. "But in drought years, we provide thirty to forty percent of their water. That's the essence of what the scheme is about: insurance."

Ninety-five percent of Australia's rice crop is grown in the Murray-Darling Basin, which holds three-quarters of the nation's irrigated farmland and provides grazing for hundreds of thousands of sheep and cattle. "I think the Snowy has worked perfectly," Dunn told me. "It has allocated the resources of electricity and water so that their benefits go back into the economy."

But the Snowy Scheme is in the middle of an escalating tug of war. "There is conflict," admits Dunn, "since what is good for water regulation is usually not good for electricity production and vice versa. The plan tends to be driven by electricity, subject to a minimum guarantee to irrigators." New claimants have complicated the conflict further. The tourist town of Jindabyne wants higher water levels year round in what the Authority thinks of as a reservoir but the town considers a recreational lake. Two towns downriver from Jindabyne—Dalgety and Orbost—want the lower part of the Snowy's natural flow restored.[129]

"This is a serious issue for us," Dunn tells me. "There are only a few hundred people altogether in the long, wild stretches of the lower Snowy River basin between Jindabyne and Orbost, but their voices have been bolstered by environmentalists." Jindabyne wants its lake, Dalgety and Orbost want more river, and the irrigators in the Riverina Plains are dependent on their water insurance in the mountains. And then there are the millions of householders with air conditioners and refrigerators, from Sydney all the way to Adelaide.

In the meantime, government treasuries have decided that the Snowy Mountains Scheme should be corporatized, which means that it must become a profitable, self-supporting corporation. The resources of the Snowy Scheme, protector against blackouts and drought, are now corporate assets as well.[130]

I spoke with Snowy Mountains Authority commissioner Murray L. Jackson about the Snowy's future. It was he who prepared the Snowy for the corporatization. Jackson is hearty and amiable but it would be a mistake to read him as merely genial. Conflicting demands on Snowy resources require more than good public relations. "When a 660 megawatt set trips out, which happens on the coal fields, then we pick up most of that demand in less than two minutes," said Jackson. "The ability to do that is critical to the assured supply of electricity to customers in Sydney, Melbourne, Adelaide, and all the places in between. That national electricity market requires us to provide continuous backup services, because we're the ones that provide spinning reserve, reactive power, and frequency control. We are the biggest risk insurance company in the national electricity market. Now, the value of those backup services will be determined by the competitive market."

Throughout the twentieth century, governments controlled almost all of the business of power generation. That is changing quickly, as deregulation moves through an energy-hungry world.[131] Competing demands can stretch water resources beyond their capacity. And what's good for the environment or perhaps a fish—full reservoirs help migrating fish over dam tops and out of the way of turbines—is not simultaneously optimal for a microchip manufacturer, a potato farmer, or a householder. As divergent pulls on the Snowy Scheme, like many others around the globe, increase, water policies will change as the businesses of electricity and water become more competitive. Water assets moving into an unregulated open market can put stream flows, rivers, water supply, and wildlife at enormous risk. In the long life of the Snowy Scheme, a great deal depends on sound political management.

As electricity contracts change along Australia's mountain-fed rivers, so does the comfort zone for irrigators who fear they will be without a voice in a corporate Snowy. Jackson seems unfazed. "Year in, year out, through droughts, floods, and famines, we deliver a minimum guarantee," he said. "That assurance gives comfort and also makes them wary about privatizing this national icon. Within the capacity of the Scheme there is tremendous flexibility to generate a lot of electricity without losing the reserve for water supply."[132]

When water shortages became visible during a drought in eastern Australia in the early nineties, the Snowy Authority hit the news again, and I heard people who didn't know the Snowy's history talking about how "the bastards stole another river." Even the best system—and the Snowy is a good one—is subject to changing climates, physical and political. Murray Jackson was sure that the Scheme would be managed for the good of all the people, but another commissioner has replaced him. A public inquiry has taken up the matter of water use and declared that 28 percent of the Snowy is to be given back to the river. At the same time, the Snowy Mountains Authority is making its way in a market economy at a time when water and electricity become more valuable by the month.

"There is no limit to the life of the Scheme," said Jackson. "The plant, like any electric motor or generator set, will from time to time require retightening of the windings, reducing the clearances on the blades, or watching for silt erosion. The success of the Scheme in the long term will be dependent on its ability to coexist with the environment."

BEFORE I WENT into the mountains one last time, I met up again with engineer Doug Price. Like all Snowy alumni, he's fierce in his pride. "It's been a great success and everybody acknowledges it," he says. He reminds me that the American Society of Civil Engineers identified

Snowy in 1969 as one of the Seven Engineering Wonders of the World and recently put a plaque on Tumut 3 Power Station, formally identifying it as an engineering landmark. Price, who left the Snowy Authority to become the managing director of the Snowy Mountains Engineering Corporation and has since worked all over the world, has little difficulty with changes affecting the Scheme. "There are discussions going on about water sharing, whether or not more water should be released for the downstream users, with modern views being brought to bear," he told me. "These considerations are proper when you consider that the Scheme was planned fifty years ago and environmental thinking now is different regarding what's proper for the use of water and for river regimes."

After talking to Price, I drove across the Snowies from Jindabyne to Talbingo and eventually reached the hillside looking over the clean sloping lines of Tumut Pond Dam, Price's first dam. Later I drove by the majestic reservoir at Talbingo with its white steel pressure pipelines climbing over the mountains, each big enough to accommodate a double-decker bus, and it seemed clear that Price had earned his pride.

Biologist Jared Diamond has written, "Australia is the continent whose human population faces the most problematic future." The Snowy Mountains Scheme is a happy combination of good geographical location and careful management. The costs of building dams have risen greatly, and today it would be impossible to build such a project as inexpensively. In a dry and thirsty land, Australia's Snowy Mountains Scheme is a valuable resource. At this moment, there isn't a better way to generate this much electricity than by storing these mountain waters in high lakes and using the drops to generate electric power—year after year after year. Over the long haul, the Snowy Mountains Scheme will pay for itself and then some. I remembered Doug Price's final words to me: "The Scheme," he said, "it has done its job and more."

HYDROPOWER NOW SUPPLIES twenty percent of the world's energy. Boosters of hydroelectricity say that we could produce five times that amount, the equivalent of about one year's worth of oil production worldwide. Can dams be made to work well enough to justify building new ones? Just a few of the difficulties of large-scale hydro projects are that they are wildly capital intensive and subject to decisions by people who profit from their building. They also destroy massive amounts of land and dislocate thousands of people, almost always the poorest of the poor. A 1997 report by the International Union for Conservation of Nature (IUCN) and the World Bank identifies the criteria critical for a hydro project to be sustainable: neither those displaced by dams nor those living downstream should suffer; waterborne diseases must not increase; river sediment has to be taken into account; there should be no loss of species or biodiversity; when a dam cuts into agricultural pro-duction, its benefits should outweigh its losses; water quality cannot be allowed to deteriorate; and the project must fit into the society, culture, and future of the region. The IUCN report suggests that the best way to reduce intolerable effects is to factor social and environmental criteria into the project analysis, which means entering all of the costs associated with them into the budget. In the past, environmental assessments have arrived too late in the planning process to have meaning. Resettlement today involves much more effort and money than has been provided in the past, and a great number of projects will be economically unviable on that basis alone. "Those deciding what projects are built, when, by whom, and how must be held accountable for the final results," reads the report.

Brazil's Itaipu Dam on the Parana River, born out of rising oil prices in the 1970s, is the world's largest hydroelectric facility. At a cost of more than $18.5 billion, it is also the most expensive, at least until China's $25-billion Three Gorges Dam is finished in 2003. Itaipu, with a capacity of 12,600 megawatts, produces as much electricity as 13 nuclear plants and saves the country roughly 350,000 barrels of oil a day and $10 billion a year. At the entrance to the dam site, a sign boasts: "Enough energy to illuminate the planet for a full month."

There are 594 large dams in Brazil, 1 percent of all the big dams in the world. While hydropower dams produce 90 percent of Brazil's electricity, the country still suffers from a shortage of power and continues to scour the Amazon Basin for dam sites. Brazil's dams have already created havoc with Amazonian jungles, opening pristine areas to logging, mining, and land speculation. There are 21 hydropower dams in various stages of construction in Brazil, including the new 1,814-megawatt capacity Sergio Motta Dam on the Parana, which has put 556,000 acres of land under water.

There is an added hazard associated with dams built in tropical forest. When a dense jungle is flooded, the rotting vegetation creates methane gases, twenty times more potent than carbon dioxide. The Balbina Dam on the Amazon is the most notorious example: Having flooded 236 square kilometers of jungle, it produces only half of the electricity needed by the city of Manaus yet generates more greenhouse gas than a coal-burning plant producing the same amount of energy. A recent study of reservoirs by the National Research Institute for Amazonia has determined that even though Balbina is an extreme example, and emissions vary greatly from dam to dam, greenhouse gases will continue to roll off those reservoirs for decades. Institute scientists believe that emissions from reservoirs may ultimately produce as much as 7 percent of the greenhouse gas contribution to earth's warming.[133] To the already daunting list of requirements for sustainable hydro development add the imperative fact that greenhouse gases from the rotting of biomass in reservoirs must not exceed those of gas-fired plants.

The emission of methane gases can be partially countered by stripping vegetation. Canadian engineer Gary Ackles has designed a cutting barge that makes it possible to harvest timber out of dam reservoirs. He expects to take $600 million worth of trees out of flooded rainforest in Brazil. The 4,000-megawatt Tucurui Dam built in the 1980s made underwater logging in its thousand-square-mile reservoir a profitable industry. Although the removal of less valuable scrub is prohibitively expensive, and some plant life was left behind, the emissions produced by Tucurui are only half those of a similar size coal-burning power plant.

In the spring of 2001, after a four-year drought had emptied the reservoirs of all its northern and southeastern dams, the Brazilian government was forced to institute emergency rationing of electricity. The country was forced to cut back as much as 20 percent of its consumption of electricity. However, a great deal of the crisis can be traced to failures in management. Brazil's transmission grids are in bad shape, and sometimes, since grid lines to dams in remote forests are impractical and expensive, they were never built at all. During the latest drought, while dams in the south of the country were full, inadequate grids made it impossible to get the power to where it was needed.

More serious is the fact that Brazil's low subsidized electric rates haven't covered the escalating costs of energy production. In the 1960s, building power projects cost only $800 per installed kilowatt. The costs of new projects have risen to between $2,000 and $6,000 per kilowatt, escalating figures rarely accounted for in customer charges. Higher power costs encourage efficiency, and if electricity were sold at its actual cost in Brazil as in other countries, fewer power plants would be needed. One study suggests demand-side management in Brazil over fifteen years would save $52 billion and 26,000 megawatts of new power plants. Writer Patrick McCully spells it out: "This is equal to the amount of power produced by more than two Itaipus—or well over a hundred Balbinas." Norbert Walter, chief economist of Deutsche Bank, says that if the United States, which uses 40 percent more energy than other developed nations, met European efficiency standards in the 1990s, it would have already surpassed the carbon-dioxide-emission standards set by the Kyoto Agreement by 22 percent.

A host of new large-scale dams along the Amazon may not be a satisfactory solution to Brazil's problems, yet until other renewable energy sources become more accessible, the alternative to hydropower at present is fossil fuel. Dozens of new gas-powered stations are in planning stages in Brazil, and a full-scale atomic program will soon add several new nuclear facilities. Had the country cleaned up porous systems and controlled its demand for energy, many new power facilities would be unnecessary Whatever they hurriedly put in place—big dams on the Amazon or fossil-fuel plants—will bring about incalculable damage. Brazil's problems are a vivid reminder that we need to bring all of our knowledge and skills to bear on solutions before we are in the heart of a predicament.[134]

IN THE FACE of rolling blackouts, when our computers crash and air conditioners and elevators fail us, we too will have to build new power plants of one kind or another. After the California energy crisis of 2001, shortages and high prices meant firing up old and dirty coal plants and subsidizing fossil-fuel technologies and nuclear programs. A chart of new megawatts now being added to American power grids has spiked higher than my electric bill.[135] Vice President Dick Cheney says that America will need to build 1,300 new power plants in the next 20 years.

Considering escalating energy demands, I asked Bill McCormick, chief engineer of Wyoming's Buffalo Bill Dam, what he thought about power from water. "I'm a firm believer in hydropower," he said to me. "If you can run that water through a turbine and then onto a barley field, it's renewable and reusable energy. You don't have to worry about disposing of nuclear waste or coal emissions. The power plant at Buffalo Bill Dam produces about 100 million kilowatt-hours of power each year—more than enough for all of the 23,000 people living in the county. It's a significant source of clean power."

However, as much as Bill McCormick believes in hydropower, he doesn't feel the best answer is more big dams. "America as a whole hasn't looked at hydropower very well," he says. "They want everything from big systems, to store water in Lake Powell, Fort Peck, or Garrison. A lot of little units would be better. The Bureau was caught up in that way of thinking. We didn't put power on projects that had small releases. But if you already have a reservoir, it's foolish to watch the dollars roll by. Now with microprocessing, you can put a hydro unit every place water falls."

McCormick explains that if you have some running water and a drop—a creek that pours 5 gallons a minute and has a fall of say, 10 feet—you can generate a considerable amount of hydropower that doesn't cost much, in any sense. You don't even need a dam if you take some of that water and run it into a penstock through turbines and back into the stream. Small or low-head hydro and micro-hydro—units producing less than a megawatt of power—are cheap, clean, and a good supplementary supply for communities of all sizes. Because it makes use of local materials and skills and is more reliable than power from government grids, small hydro can be acutely important in developing countries and in remote areas. Costs vary widely, but micro-hydro is competitive even in remote places when taking into account the cost of grid connections or the use of diesel fuel in generators. It has already been successful in Nepal, in the tea estates of Sri Lanka, and in China. In the United States, the U.S. Department of Energy says that we can easily squeeze 10,000 megawatts out of small-scale hydro—enough power to serve about 10 million people.

Even more cost-effective is increasing the capacity of existing dams. Brazil is adding two new turbines to Itaipu, which will supply an additional fourteen hundred megawatts—enough electricity for 1.5 million people—at a cost of $185 million (a comparable hydroplant would cost upward of $1 billion). In the United States, the Bureau of Reclamation not long ago added sixteen hundred megawatts to several of its dams over a decade at a cost of $69 per installed kilowatt. New project costs run nearly a hundred times higher. But the biggest bargain in power production is retrofitting. Most American dams were built for mills, flood control, or irrigation. Just 3 percent of them produce electricity. The U.S. Department of Energy has said that as many as 2,600 existing dams can be retrofitted for hydropower production.

A New Englander named John Webster has shown how to make both small hydro and retrofitting work. A resourceful farmboy from Maine, he trained as a steam-and-diesel engineer and sailed the world for ten years on Gulf Oil tankers. When he decided to marry and get off the seas, over a decade ago, he began to investigate the possibilities of small hydro in a number of small, decommissioned dams with powerhouses in New England. After a few bruising encounters with town councils—he says that water politics are fierce even at municipal levels—he was able to buy an old granite arch dam near South Berwick in New Hampshire. He poured a million dollars into revamping the dam and its powerhouse; installed three Swedish submersible pump turbines; and began pumping 3.5 million kilowatts of electricity a year. He sells that electricity to the local utility at a bargain rate of .09¢ a kilowatt-hour and makes a profit of roughly $300,000 a year on his investment. Webster has successfully revamped and added power capacity to three other projects, including an abandoned sixty-foot-long arch dam with a drop of 200 feet at Minnewawa, New Hampshire, near Marlborough.

Webster's installations are all inconspicuous, the dams small and the drops pretty, as anyone who has ever driven through the old mill towns of New England knows. Webster is now looking at small projects in South and Central America. "There is an especially great need for hydropower in developing countries," he told me with obvious pleasure. "Hydropower gets knocked on the head while energy marketers are building large gas-fired plants that pump sulfates and carbon dioxide emissions with no concern about global warming. In New Hampshire alone there are forty megawatts of small hydro and more in Maine. We have so much going for us."

IN THE LAST FEW YEARS, we've all read in the papers of efforts to tear down old dams around the country in order to restore fish populations and wild and scenic rivers. To great acclaim, politicians turn up wielding a ceremonial sledgehammer and, after some explosives are set off, a river goes free. More than 460 dams have come down in the United States in the last decade, but environmental groups like American Rivers point out lustily that there are over 2.5 million still standing.

The Kennebec Coalition, an organization of conservation groups that has successfully lobbied to get the 1837 Edwards Dam on the Kennebec River torn down, came up with three questions that can be usefully asked about any dam: Why is it here? Can the harm it does be reversed? Do its benefits exceed its costs? These are good questions. In the case of the destruction of the Edwards Dam, because it was close to the coast, nine sea-run fish species, including the shortnose and Atlantic sturgeons, are now able to reach upstream spawning grounds. Atlantic salmon and striped bass promptly began to do so. At the same time the Edwards Dam came down, however, another environmental coalition, the Allagash Alliance, determined that the old Churchill Dam in Maine was important to the preservation of the Allagash Wilderness Waterway and argued to have it replaced by a newer dam. These apparently contradictory decisions by members of the environmental community signal a welcome sophistication in the understanding of different conditions. Both groups were right.

The questions put forward by the Kennebec Coalition ought to be asked about each of the hundreds of federally regulated dams up for renewal in America in the next decade. Some dams ought to be dismantled. Others ought to remain and perhaps produce electricity. The answers are not always as obvious as they seem. A dam shouldn't come down just because it's there, any more than a river should be dammed simply because it can be. Using capacity that we already have could prevent the building of some coal and gas plants and buy us time until solar power, wind power, and other renewables become significant parts of our grids.

I spoke to another electricity expert, S. David Freeman, former chairman of the Tennessee Valley Authority, about the role of hydropower in easing the escalating energy crisis. "People forget that when the TVA electricity came in, it replaced a coal-fired generation," he said. "It is a very powerful clean-air partner."

Despite his faith in hydropower, when Freeman visited China in 1980 he strongly urged the Chinese government not to build the monumental Three Gorges Dam, which, at a capacity of 18,000 megawatts—the equivalent of burning 40 million tons of coal a year—will be the biggest hydro facility in the world. Instead, Freeman told the Chinese that cascades of small hydro dams on tributaries could be built more quickly and cheaply and could get more out of their rivers without doing as much damage to the environment. But he understands China's need for power. "Harnessing the rivers of China may be the only answer to global warming; otherwise the Chinese will burn all that coal," says Freeman. "Knee-jerk opposition to any water power project anywhere on earth is as environmentally destructive as it is deeply destructive to build anything, anywhere, anytime."

"Should India build a new hydro-plant, displacing thousands from their homes, or a dozen more coal-fired power-stations to pollute the atmosphere?" asks Janet Ramage, author of *Energy, A Guidebook*. "Should the Swiss drown a few more Alpine valleys or construct another nuclear power plant? . . . Should I replace my ancient boiler or do something about the draughty windows?" Neither the Snowy Mountains Authority nor the Tennessee Valley Authority could provide enough inexpensive power to meet the burgeoning demand for cheap electricity when people began to fill up the landscape. Hydropower will never be a complete answer to the world's energy problems or global warming, but it can be helpful.

In England, for example, there is a plan to build a barrage across the Severn Estuary and outfit it with 200 large turbines that would provide 7 percent of the entire country's electricity—the equivalent of 12 nuclear power stations. Small hydro installations on the pulsing drops along India's canal systems could supply hundreds of towns and cities with all the light and refrigeration they need. Run-of-River schemes, which divert water and return it without reservoirs, are possible in many sites. The most important use of dams may soon lie in the fact that the large amounts of electricity needed to produce hydrogen for fuel cells can easily be generated from freely cascading water and stored. Iceland's efforts to do away with fossil fuels and become environmentally benign is possible only because it has such immense hydroelectric potential in its fast-running rivers, which will be used to produce and store hydrogen for use in all its vehicles and power stations.

Engineer Samuel C. Florman writes: "Like electric power in Brazil, our technological predicaments can be so confounding that they force us to stop in our tracks and reflect on who we are and what we really want." We all seem to want electricity. Meanwhile, the Kyoto Agreement, at which the United States continues to balk, asks for a reduction in fossil-fuel emissions of a mere 5.2 percent. The scientists of the Intergovernmental Panel on Climate Change—identified by *The Economist* as occupying the "middle ground" in the debate—want us to cut emissions by 50 to 70 percent. Difficult choices are ahead of us. Not many will give up electricity, but we have to be realistic about where it comes from, even when power plants are not puffing carbon dioxide and sulfates in our sightlines.

I've recently read about a family in Texas that has a small hydro installation on their stream, solar energy collectors on their roof, and a wind pylon in the backyard. I wish I could be like them. But I live in a city. I don't have a stream or a backyard, and my water comes from an upstate reservoir and is stored in a tank on the roof above my head. If I'm going to be more intelligently self-sufficient, I'm going to have to rely on my political representatives to improve their understanding of the mix of energy sources we all need.

Chapter 6

RAGING RIVERS

Living in Floodplains

Floods are acts of God. Flood losses are largely acts of men.

GILBERT F. WHITE, *HUMAN ADJUSTMENTS TO FLOODS,* 1945

One of the most breathtaking hours of my life was spent in an airplane following the broad, silvery Brahmaputra River out of the Himalayas, across Assam and Bangladesh, into the floodplains where the Brahmaputra meets the Ganges and the Meghna in a wondrous, terrible mingling. The world's biggest mountains were pink in the morning light and seemed docile as we flew down the wintry river basin, but I knew it wouldn't be long before snowmelt and monsoon rains would fill the rivers to overflowing and send their waters south in murderous rage.

The delta of Bangladesh is the largest and most intensely used delta on earth. Assaulted from the north by raging rivers and battered by cyclones blowing in off the Bay of Bengal to the south, it is the most treacherous and one of the most volatile on the planet. On a single night in 1970, a storm took the lives of 225,000 people. Twenty-one years later, in 1991, a twenty-foot wall of water traveling at 145 miles an hour washed over two-story buildings as if they were mailboxes, and killed 138,000 more.

Bangladesh is beautiful, however, and so are its rivers—land banked high around their edges, the surrounding fields deeply green with crops, broad waters dotted with boats easing into the mouths of the Ganges. Three great rivers twist together across the plains, braided like the plaits worn by the women on their banks. Intricately wound with dozens of smaller rivers, they form a watercourse two and a half times broader than the Mississippi, having absorbed four times as much rain in the jungles and mountains above them.

Rivers wander. In 1966, the Ganges and the Brahmaputra, joined in their lowest reaches, shifted laterally northward almost a mile, gouging a new channel 100 feet deep. Twenty-two years later, the river jumped again, this time only half a mile away, but deeper, 150 feet down. The Brahmaputra, brooding and dangerous, as much as 30 miles wide in flood, moves over a self-sculpted alluvial fan and may soon move again. These rivers shift in their multitude of channels so often that maps of the delta illustrate a landscape as wriggly as a basket of snakes.[136]

I asked a young man from Bangladesh's capital, Dhaka, if his city ever went under water in time of floods. "Madame, the whole country is under water in flood." He laughed. In flood season, I soon learned, anywhere from one to two thirds of Bangladesh goes underwater. In the last fifty years, the Brahmaputra has flooded forty-seven times.

I FIRST LEARNED about Bangladesh far away from its crowded riverbanks, in a place that nevertheless affects them, the quiet streets of Den Haag in the Netherlands. There, I met a young engineer, Gijsbert Te Slaa, senior consultant in hydraulics and coastal engineering in the headquarters of the Dutch engineering consortium NEDECO. Sandy-haired, earnest, good-natured, Te Slaa, like many of his countrymen, became an engineer out of tradition. Te Slaa has built ports and bridges in Malaysia, India, and Bangladesh for a NEDECO firm, Royal Haskoning, and he has worked on the tricky business of controlling rivers as well. "River engineering is a far different kettle of fish from working on a port," he tells me. "It is more complex. The human factor plays a much larger role and nature is less predictable, particularly in terms of floods."

The chaos of Bangladesh—which Te Slaa confesses, he loves—is a far cry from the orderly cities of the Netherlands, but in at least one respect the two countries are similar. "Bangladesh, like the Netherlands, has been formed by the deposit of silt from the mountains into the sea," he explains. "Land is emerging, but at a pace you'd hardly notice. The mountains are broken down by the sun, ice, and water, and the sediment is transported by rivers to the sea. Initially the gradient of the river is steep and the capacity to transport sediment is high, but as the river approaches the sea, the gradient diminishes and sediment is deposited in the delta, very, very slowly. Still the water comes, in some years more than others, and so, from time to time, there is flooding."

The same floods that take such a fearful human toll in Bangladesh are indispensable to the lives of those that survive them. Each summer, the rivers rush south and replenish groundwater; nurture crops and fisheries, which are the country's source of protein and 10 percent of its economy; and bear loads of silt that renew the soil.

"When you see the rivers, you are inclined to think only of water," says Te Slaa. "But the river also delivers sediment, an important element in the formation of the beds of the river." The silt is vital to agriculture, but it also dangerously alters Bangladesh's land and waterscapes. "Sediment erosion or deposition works at a different time scale than the river," says Te Slaa. "If you narrow a river, or if you deepen it, or if you extract water for irrigation, you diminish the flow of the river immediately, but the sediment is not changed. That has irreversible consequences that may only become apparent over a very long time."

The Ganges, Brahmaputra, and Meghna, pouring out of the highest, most earthquake-prone mountains in the world, carry more than 2 billion tons of silt every year into the shifting, sliding delta, raising the rivers higher and higher on their beds. "It is purely nature," Te Slaa said to me, "just as it has been in the delta for thousands of years. Since people have interfered with the sediment transport capacity of the river by building protection works, the riverbed is now in some places higher than the original land and the embankments or dikes have to be continually raised to keep pace."

The large system of dikes along the rivers of Bangladesh meant to protect the populace have instead made the region less stable and more vulnerable to floods and storms. Upstream embankments keep water in the river that would otherwise have spilled into floodplains, amplifying downstream flows so that those living inside embankments are in ever-greater danger when floods overtop the dikes. Embankments that keep water away from crops in normal seasons trap excess water during floods and often have to be broken to let water out. Embankments deprive land of silt, so that farmers have to rely on expensive fertilizers, and often come between fish and their spawning grounds. Finally, the 4,000 miles of earthworks in Bangladesh—saturated and unstable— need constant reinforcement and maintenance, a nearly impossible task for one of the poorest countries in the world.

The massive sediment loads carried by the Brahmaputra into the embanked, moving world of the delta, are the second largest of any river sediment tonnage in the world, pushing ever farther into the Bay of Bengal. The sediment, emerging then disappearing, brings a kind of precarious fortune to the poor people who live here in the form of *chars,* shifting bits of emerging silt that push up from the water only to vanish later. One such island, Urirchar, a full eight miles square, rose up out of the eastern delta in the late eighties. Although the Bangladeshi government formally forbids settlement on these little islands until mangroves have been planted to stabilize them and embankments and cyclone shelters have been constructed for protection, the large numbers of people in the delta, many displaced by other floods, are quick to seize even the smallest piece of ground. Urirchar was quickly taken over by farmers desperate to get a crop of rice into even questionable ground.

There are thousands of *chars* in the mouths of the Ganges. As rivers, storms, and floods move through the landscape, the *chars* often disappear as quickly as they arrived, along with the squatters and livestock and thatch houses that occupy them. Sometimes landowners will continue to pay property taxes on *chars* that have gone underwater, in case they emerge. In an enormous summer storm in 1985, Urirchar, like many little islands, was wiped clean, its inhabitants scattered like twigs. Many drowned. But the rest returned as soon as the clouds had passed. They had nowhere else to go.

Population is the country's single most pressing problem. One hundred and thirty million people live in Bangladesh, a country the size of Wisconsin, and half of that population lives in the uncertain landscape of the rivers. Although growth has slowed in the last decade, thanks largely to nongovernmental organizations that have encouraged birth control and prompted the empowerment of women, little hope exists for ending the progression and enormity of disasters here unless the population problem is solved.

It is unthinkable to abandon the people who live along these perilous rivers, yet difficult to know what to do about them. India, in search of water and hydropower for itself, has proposed building six mega-dams upstream in Assam and a link canal long enough to carry water out of the Brahmaputra River across Bangladesh to the Ganges. On paper it seems a plausible scheme. By lessening the amount of water in the rivers, it would also lessen the severity of flooding. But the ecological costs would be tremendous. The priceless forested floodplains of Assam, the last stronghold of the Indian rhinoceros and home to perhaps the richest wildlife population in Asia, depend on the regular inundation of the Brahmaputra and absorb important quantities of the river's overflow. To deprive them of the annual floods would be to destroy them.

The engineering projects themselves are at best questionable. The worst of them is the link canal, which would harm both wildlife sanctuaries and the intensely packed floodplain farmlands, and would be impossible to drain in the flat, low-lying plains. The dams are dubious as well. Reservoirs would perch behind dams in seismically tremulous regions. Because the Brahmaputra is among the siltiest of rivers, the storage of the first mega-dam is predicted to be filled by the time the last one is finished. Nor is it likely that Bangladesh would feel comfortable with the control of the Brahmaputra in the hands of her neighbor. Although the two countries share a simple treaty over the Ganges, India's record with the Farakka Barrage on the Ganges has historically been to remove water for irrigation in the dry season yet give it full flow in flood.[137] "Perhaps they will build dams to create reservoirs in the Himalayas," Te Slaa said worriedly. "I'm very afraid of what it will do to Bangladesh.

"You have to look at all the effects, not only the water but also sediment," he reminded me. "What are *all* the consequences if I interfere with nature? Not only short-term but long-term. When you are planning to do something about a river, you have to look far into the future to think about consequences."

It's asking a lot for a country like Bangladesh—struggling simply to keep up with the needs of its people—to look very far ahead. After floods in 1988 swamped unusually large quarters of Dhaka, the government of Bangladesh called for international help to build even more, bigger, taller, stronger embankments. The World Bank and fifteen donor countries, including the Netherlands, met to begin planning flood protection. In 1990, out of this effort evolved a strategy, *Flood Control in Bangladesh, A Plan for Action*. The plan, largely structural, would cost between $1 billion and $7 billion and comprise the world's largest single aid project.

But the plan has been hotly debated both in and outside of the international donor community and among the increasing numbers of people who give their attention to water issues. American water experts Peter Rogers, David Seckler, and Peter Lydon have submitted a counterstudy of the floodplains of the eastern rivers to the World Bank, pointing out that dikes that would cost $6 billion to build would also require $6 million to maintain every year. They recommended that donors help people live with floods rather than keep up the relentless fight to prevent them. Since the people of Bangladesh have already become ingenious at surviving floods, Rogers, Seckler, and Lydon said that what they really need is help in doing so.[138]

In the delta of Bangladesh, farmers long ago adapted to their floating world, living on high mounds and growing a fast-developing strain of rice to beat the floods. The American experts' report to the World Bank, titled *Eastern Waters Study*, says that accommodation means that for people to stay, some flooding must be accepted. This means using strategies to mitigate floods, such as converting farms to fish ponds and offering relief from hazards to those who remain; plenty of warning, places to go that are stronger and safer than mud huts, and the means to get there. They need to be assured of rescue, food, and medical care; more than half of flood-related deaths are from diseases—cholera, dysentery, or malaria. In the disaster of 1991, on the island of Sonadia, 650 people survived because they were able to reach shelters. Thirty years earlier, there was no shelter, and the entire population of the island was lost in the floods.[139] The great storms that come out of the Bay of Bengal cannot be stopped, and the silt load brought by the Brahmaputra is too much to dream of dredging. The forces of nature along this multitude of rivers will not be easily resisted. But along other rivers, where warning systems are better and evacuation more efficient, the loss of lives is smaller.

When I spoke with Gijsbert Te Slaa, he had just returned from the Brahmaputra north of Dhaka, to unite a part of the country sliced in two by a river that has historically been anywhere from five to fifty miles wide. The people of Bangladesh have wished for a bridge here for many years, but even a bridge over such a changeable river is a weighty task. "Sooner or later the river will change position," Te Slaa told me. "In this floodplain, two, three, or four channels continuously change position. To cross all the channels, you can make a very expensive bridge—ten miles long—and wherever the channels go they will always be under the bridge. But if you can arrange a corridor for the water by using training works to keep the rivers in their channels, the bridge only needs to be three miles long—or five miles in case of a severe flood with no undesirable side effects—and you have saved two-thirds of the cost. I was brought in to research whether or not it can be accomplished."

Te Slaa found that it could be done. His bridge is finished, and it is a wonder. The training works—slopes, aprons, and guide bunds—to keep the river in place are protected by blankets of geotextiles, asphalt revetments, and 1.5 million tons of stone, expensive but useful techniques learned in the Dutch Delta.

"After every flood, the question arises whether or not the money would be better spent on flood control," says Te Slaa, "but I think that this bridge is a good project. Electricity extends across the new bridge, markets and producers are connected, and the isolation of one part of the country from the other has ended. Engineers can find some satisfaction in making better the lives of those that live there."

But the young engineer has no illusions about the permanence of his efforts. "In Bangladesh, like a drop of water on a hot plate, the effects of the work of engineers evaporate instantly," Te Slaa told me. "There are still a hundred-and-thirty million people living in a land that can support thirty or forty million."

When the annual floods come, foreign reporters ask the people why they stay. The answers never vary from year to year or *char* to *char*: "Yes, they told us to go. But where would I go? How would I go?"

This riverine land is not a safe place to live, but its people have little choice. In the United States, people have chosen to encroach on unsafe places. Americans, however, expect their government to help them stay there.

THE MISSISSIPPI RIVER, fed by the Missouri, Minnesota, Black, and Illinois rivers, stretches 3,710 miles—only the Nile, Amazon, and Yangtze are longer. It is also America's most flood-prone and unpredictable river. Mark Twain spoke of the Mississippi River's "disposition to make prodigious jumps by cutting through narrow necks of land, thus straightening and shortening itself. More than once it shortened itself by thirty miles at a single jump! These cut-offs have had curious effects: They have thrown several river towns out into the rural districts, and built up sand bars and forests in front of them. . . . Such a thing, happening in the upper river In the old times, could have transferred a slave from Missouri to Illinois and made a free man of him. . . .

"When the water begins to flow through one of those ditches I have been speaking of," wrote Twain, "it is time for the people thereabouts to move."

But they don't. Instead, at the hands of engineers attempting to tame its unruly nature, the Mississippi has been pinched, stuffed, and leveed to a fare-thee-well. As early as 1914, William Willcocks chided American engineers for their work on the Mississippi's banks. "The levees are too long and everybody knows it," he barked. "The conditions on the lower Mississippi today are very serious and delay may produce a flood, which may be not unlike Noah's Flood." Willcocks didn't doubt the objectives of flood control but he was aghast at the backward way in which Americans were going about it. "The Mississippi has been reclaimed, unfortunately for itself, from the mouth up," he told his American audience. "Louisiana made for itself fields and gardens behind its levees. When the next state farther up the river made its levees and concentrated the flow, it put Louisiana into difficulties. Louisiana raised its levees and then it went easier until the next state upstream started to build. And however hard Louisiana works, it can barely keep ahead of the difficulties brought on it."[140] Willcocks was right—New Orleans barely survived the flood of 1927. Its survival was largely due to the drastic fracturing of levees upstream.

In America, we don't think much about rivers until they overrun the land. When they do, the numbers of people lost are usually smaller than in poorer countries but the value losses are far higher. The Army Corps of Engineers has built more than ten thousand miles of levees and flood-walls across the country. With our rivers leveed, dammed, and seemingly well-protected, floods have largely lost their menace. Even after a river has flooded, in a few years memory fades and back we go again, with bulldozers and backhoes, to dig and build another time. Eventually, some river or another will rise up to remind us of our foolishness. At the end of the twentieth century in the United States, ten million homes were planted solidly in the wandering zones of rivers, and floods were the number-one destroyer of property and lives.

American flood policy has historically been prodded by hydrological events on the Mississippi. Until 1927, there was a "levees only" protection policy along the river, and no levee built to Mississippi Commission standards had ever broken. But that year, levees broke in large numbers as the bursting river spread itself a hundred miles wide across the basin, reclaimed 20,000 square miles of its floodplain, and killed 313 people. After it was over, the government became even more ferociously committed to protection and passed a series of congressional acts to pay out more and more damages for flood losses. In the 50 years following the flood of 1927, the Army Corps of Engineers' response to floods was to build dams, levees, and floodwalls.[141] In spite of all those embankments and all that concrete, losses mounted steadily. In the 1960s, the National Flood Insurance Program was created to soften the burden on the general public. But only a few people bought the insurance, relying instead on luck, God, or, most reliable of all, disaster relief.

In 1993, in spite of 7,500 levees flanking the Mississippi system, the event that came to be known as the "Great Flood" tore into nine states and inflicted $18 billion worth of destruction. "You farm down here, the river will find you," Art Heinicki, an Illinois farmer said to a *New York Times* reporter. Large quantities of oil, sewage, pesticides, fertilizer, and the hazardous wastes from 54 Superfund sites washed downstream. Fourteen hundred private levees were either breached or damaged. An indeterminate number of government levees collapsed as well. The federal government shelled out $6 billion and the states another $1 billion to pick up the pieces after the disaster. "We're in a pickle," said Mark S. Alvey of the Army Corps of Engineers.

The Flood of 1993 was a horror even by Mississippi River standards. For duration alone it was remarkable—rain fell heavily along the lower Missouri and upper Mississippi for almost a year prior to the full catastrophe, and this was followed by exceptionally heavy snowmelt. All of this meant that by May of 1993, lakes and reservoirs were full to capacity, the soil was saturated, and there were flash floods. It continued to rain so persistently that a kind of atmospheric gridlock set in, a self-perpetuating cycle in which water evaporated and fell, over and over again.

By mid-June the Missouri and other upper Mississippi tributaries began to flood, and in the last week of June the Mississippi itself reached flood stage. By early July, levees were beginning to break—120 on the Mississippi alone. In mid-July the National Guard began evacuating and hauling sandbags onto levees. Nevertheless, the embankments began to break south of St. Louis, and 8 million acres of cropland went under water. In early August, still more levees caved in. At last, the flood crested, and by August 9 river levels everywhere began to fall. Even so, flash floods continued throughout the month, and rains didn't stop until the following year.

For the Missouri and upper Mississippi region, it was the most expensive flood on record. President Clinton declared 525 counties in nine states as disaster areas. In early 1994, while it was still raining, the president's Floodplain Task Force established a review committee to pinpoint the disaster's causes and consequences. The committee's report suggested that while there still were places where structural control was advisable, protection was no longer the sole answer to floods. The committee proposed a combination of solutions, such as restoring wetlands to absorb flooding, and encouraging or even sometimes requiring people to keep out of floodplains.[142]

A LEVEE IS SIMPLY what Americans call a "dike." It is typically a substantial embankment—wide and solid, grown over with grass, placed well away from the threatening river—and is almost always a mixed blessing. According to the Federal Emergency Management Agency, overtopping or levee failure causes one-third of U.S. flood disasters.

When a river is cut off from its natural floodplain, the water has

nowhere to go but downstream, searching for the weakest wall and eat-
ing at the base of its artificial constraints. A levee on the right side of a
river will increase the force of the water against the left side. Build levees along both sides and the river is channeled higher and faster, barreling down its new raceway with accelerated force. Although strong levees
may protect one town, another in a more vulnerable position will then
be placed at greater risk. St. Louis survived the flood of 1973 because
levees above it failed. St. Louis was saved for a second time in the flood
of 1993, not simply by its own levees but because levees failed south of
the city where the brunt of the river's force released itself on some
other sorry towns. The 1993 flood discharged the same quantities of
water as the 1903 flood, but, because of the extensive system of levees,
the waters rose twelve feet higher. When a levee as tall as a three- or
four-story building breaches, the water explodes with the percussive
force of a burst dam. So, while man-made channels and walls offer protection from smaller floods, they make big ones larger and meaner. And
levees are expensive. Maintenance of levees along the lower Mississippi
can run as high as a million dollars per river mile. Spending billions for
blanket rebuilding of levees and dams subsequently has lost its across-
the-board appeal.

William Willcocks said in a 1914 lecture about the Mississippi River
that before any levee is built, the height of the flood that will be artificially produced by the levee should be calculated. The height of that
flood should then be publicly displayed, he said, "so that every man could
see for himself the terrible condition which is coming."

When the town of Valmeyer, Illinois, on the Mississippi, went under
water in August and September of 1993, its people felt as if they'd had
enough. The great-grandfathers of the 900 Valmeyer farmers who had
planted their wheat along the Mississippi in the nineteenth century
had only camped out in the floodplain and then retreated to homes on
higher ground. Their less cautious descendants built their houses closer
to the river. After an occasional flood, they'd dig out of the mud and go

back to life as usual. But on September 9, 1993, the residents of Valmeyer took a radical step. They voted to move to a bluff—a mile and a half away and 350 feet up—well out of the river's way.

More and more people are catching on to the idea that when facing a major force of nature, the best option may be to step aside. Soldier's Grove is a hundred-year-old town on the Kickapoo River in Wisconsin. Around the turn of the century, the effects of logging and land clearing began to energize flooding on the Kickapoo. The first flood arrived in 1907 and returned about once every decade after that. The Army Corps of Engineers offered Soldier's Grove a dam and a $3.5 million protective levee in 1964. The town fathers figured out that if the government used the money to relocate them instead of building flood control, since they wouldn't be flooded again and there would be no levees requiring maintenance, they could repay the government seventy years sooner than for the proposed dam and levee.[143]

When the government refused to pay for relocation, the townspeople planned their own getaway. They purchased land on higher ground, planned a new city center, and ran sewer lines just out of the river's way. No mechanism existed that would allow the government to help, so small state planning grants were used to execute the relocation plan. While the planning was still going on, the river flooded Soldier's Grove again in 1978, and once again the town was offered rebuilding money in the old location. Soldier's Grove refused, citing an executive order penned by President Jimmy Carter in 1977 that disallowed federal support for floodplain development unless there was "no reasonable option." The government finally gave Soldier's Grove a million dollars toward relocation. The townspeople made their old town center into a riverside park and elevated buildings on the fringe of flood-prone areas. They put their new business and community buildings in the flood-safe zone and added solar heating. The town had at once flood-proofed and vastly improved Soldier's Grove, which became a bustling, attractive city. Property values quickly doubled.

Other communities have found innovative ways of coping with encroaching waters. After a flood drowned fifteen people and racked up $180 million in damages in the Mingo Creek floodplain in Tulsa, Oklahoma, the Army Corps of Engineers created a protective series of lakes connected by a system of ball parks, greenbelts, and jogging trails. Near Boston, the Corps again collaborated with local governments to protect 8,500 acres of wetlands along the Charles River at a cost of $10 million, instead of building dams and levees that would have cost $100 million.[144]

The federal government, having paid again and again to rebuild houses, farms, and businesses, has finally, ever so gently, begun to question the wisdom of living where a river will inevitably invade the household. The flood of 1993 prompted smarter, long-term solutions to flooding. Changes made during the Clinton administration make it easier for disaster aid to arrive in the form of buyouts, turning land to less dangerous or less costly uses, or making those who choose to live in frequently flooded properties more responsible for their own well-being. The federal government had paid out $48 million in insurance claims in 1993, having only received paid premiums of $625,000. Consequently, they toughened up flood-insurance regulations by making insurance mandatory for anyone taking out a mortgage and prohibiting policy cancellation later on. Today, flood insurance must be purchased thirty days before a flood strikes, as opposed to the old law requiring five days—a policy under whose gracious umbrella 7,800 farmers and homeowners rushed out to buy insurance when they heard that the Missouri was flooding upstream in 1993.

These policy changes have already begun to save the federal government substantial sums. The $100 million in grants allotted by the Federal Emergency Management Agency after the flood of 1993 for the purchase of vulnerable properties in the state of Missouri alone will save $200 million over a twenty-year period. By 1997, more than 200 towns had moved out of the Mississippi floodplain. Many more cling to their old homes, but attitudes along the river have begun to change.

Floods, a natural way for a river to cleanse and rejuvenate itself and its surroundings, come more frequently, now that the wetlands that once absorbed them have been replaced by cement. "There is no reservoir that you can find for a river so good as one of those open basins on its banks," said William Willcocks. Much of the Mississippi Valley was once intermittently flooded woodland and marshes, biological filters that protected the watershed and provided escapes for floodwaters— natural sponges to soak up excess liquid. The valley is now filled with cities and farms, and 85 percent of the Mississippi's wetlands have been lost, taking flood escapes and natural water filters along with them. In 1994, a $68 million allocation for flood relief to the U.S. Soil Conservation Service was set aside to acquire 85,000 acres of land to be converted back to wetlands. This enabled buyouts for farmers who, although no longer able to build on their lands, could still hunt or fish there. One happy environmentalist called it "green pork." The Army Corps of Engineers also set aside roughly 14,000 acres in Missouri for wildlife habitat, and the U.S. Fish and Wildlife Service has identified 60,000 acres of land that will become the Big Muddy National Fish and Wildlife Refuge. These wetlands will help soak up smaller floods, filter water, and create recreational and wildlife areas. Best of all, they will keep people out of the floodplain.

There are other important ideas at work for those who retain land in areas of recurrent flooding. The Federal Agriculture Improvement and Reform Act has removed incentives to grow easily damaged crops, like corn, and encourages more flood-resistant crops, such as trees or hay. The National Oceanic and Atmospheric Administration is adding Doppler radar for better rainfall measurements and more accurate warning systems, which will allow people to know what's coming in time to get away.[145] One construction company in Saint Charles County has elevated more than 120 homes so that floodwaters can pass underneath them. These houses are placed on independent, floating, deep foundations with breakaway walls. Their ground-floor walls lift on hinges out of the way of rushing water, yet the superstructure is strong enough to withstand its ferocious pounding.

THE MAN MOST RESPONSIBLE for changing the questions asked about living and building levees in floodplains is geographer Gilbert F. White, a slim, soft-spoken Quaker gentleman whose dissertation, *Human Adjustment to Floods,* first made the case for flood management rather than control. In 1966, White pointed out that since national flood-control policy was set in the 1930s, at least $4 billion had been spent for structural flood control. Yet economic losses had risen as steadily as the Mississippi in spring. White also observed that new development in floodplains was taking place at a rate comparable to the speed at which engineering works were being built to protect old development. "Only slight attention has been paid to economic justification for works constructed," he wrote.[146]

Engineers were not solely responsible, in White's view, for the extravagant bill the nation has paid to keep people living along rivers. "The engineers saw themselves as valiant competent technicians who set out to curb a stream on the rampage and to harness a recalcitrant nature," he wrote. For centuries, he says, floods have been viewed as "great natural adversaries," "watery marauders" against which men waged a bitter battle. He believes that better answers to flood losses are to be found within a wider series of adjustments. Such adaptations include flood abatement using erosion control and forest cover to hold water in the soil; the acceptance of some losses; some conventional engineering works; accurate data and forecasting; true economic appraisal; watershed planning on a basin-wide basis; but most of all, public policy that intelligently regulates the use of floodplains.

Some thinking by policy makers has begun to swing Gilbert White's way. However, the costs of changes are high and savings usually occur well after a congressman has left his post, which means there isn't much political gain in fighting to improve policy. The Bush administration has worked hard to undo efforts made by the Clinton administration to encourage farmers to change their farm acreage into wetlands, has relaxed rules on filling streams, and has even eliminated some restrictions on floodplain development. The city of Davenport, Iowa, had decided to build parks instead of levees, and as a consequence, damage in Davenport has been considerably lower than in cities that allow development along levees. After the spring floods of 2001, Davenport was criticized by President Bush's Federal Emergency Management Agency chief, Joseph Allbaugh, for not building floodwalls.

William Willcocks thought those responsible for building the control works perhaps ought to share the risk. "Imagine for an instant the senators and representatives in Washington sitting down and looking academically at the question of the Mississippi River levees," he wrote. "Or imagine, as you have here, the commission, which looks after the Mississippi levees, taking care to locate itself well up at St. Louis, where it is totally out of danger. In Babylonia they would have made them all live behind the worst of their levees and whatever else they did they would not have any breaches." Even after the great flood of 1993, our politicians haven't seemed to fully grasp the conditions with which we are dealing. Today there are substantial levee-building projects under way along the Missouri and Mississippi rivers, including one stunning $58 million project being built to shelter a strip mall. "Four years at West Point and plenty of books and schooling will learn a man a good deal, I reckon," wrote Mark Twain, "but it won't learn him the river."

Twain was talking about the Mississippi, of course. "You turn one of those little European rivers over to this Commission, with its hard bottom and clear water, and it would just be a holiday job for them to wall it, and pile it, and dike it, and tame it down, and boss it around, and make it go wherever they wanted it to, and stay where they put it, and do just as they said, every time. But this ain't that kind of a river."

New ideas about comprehensive river-basin management being put forth by engineers and men of foresight like Gilbert White have triggered changes that are making their way onto our riverine lands, however slowly. The answers to our floodplain dilemmas will lie in a combination of informed strategies—protection, adaptation, and, sometimes, simple surrender to the river's mighty ways. The city of St. Louis cannot be picked up and moved to a bluff, and I don't think many people would be willing to let New Orleans drown. But the costs of living in many places along rivers are too high, and we ought not to build another St. Louis in such a dubious setting. We've been dealing with our capricious rivers for only two centuries or so. Some of those who have faced the same struggle for millennia elsewhere in the world seem to have learned even less.

WHEN THE MISSISSIPPI poured over its banks in 1993 and drowned 52 persons, it was called the Great Flood of '93. The following year in China, more than 4,000 people died in floods along the Pearl River. In 1995, 1,200 died on the Yangtze. The year after that, the Yangtze took 1,179 more lives. No one bothered to call these floods "great," because in China nobody knows what the next year may hold.

The history of life in China's river basins is a cycle of droughts and drownings. The Chinese have struggled to control their rivers from the very beginning, through "works so ancient," Arnold Toynbee wrote, "that the memory of them was lost in the fog of legend." There are already 83,000 dams in the country—and the Chinese show no signs of stopping. Some ninety new dams are under way, among them the largest on earth. Chinese engineers intend to throttle and divert water to a degree not yet seen anywhere.

Chinese civilization was incubated in an unlikely womb, the basin of the single most deadly killer on the surface of the earth, the murderous Yellow River. The Yellow is sometimes called "China's Sorrow," and the name suits. In the past 2,000 years, it has overflowed its banks 1,500 times. In 1931—in the most terrible flood in human history—as many as 3.7 million people died beneath its murky waters.

The Yellow River falls out of the Kunlun Mountains and writhes through northern China, heavy with a yellow-ochre silt that gives the river both its color and its name. The silt, called "loess," is a powder-fine, weather-beaten soil carried by the river and by mountain winds so that it covers everything on the North China Plain. The loess earth is hundreds of feet deep, a thousand in some places, and it fills every crevice and canyon. Across this inland alluvial delta, the world's largest, the river makes its way, depositing more and more silt, raising the river bed ever higher, so that thirty feet above the plain it resembles an aque-duct more than a river, sometimes abruptly changing course. In 1851 it tore itself a brand-new channel across the country, pouring into the Yel-low Sea, 250 miles from its former mouth.

When the Yellow inevitably breaches its dikes, the waters burst vio-lently onto the flat land below, consuming everything within its long reach and endangering the more than 110 million people that live along its banks. In 1887, when the river broke its constraints, it flooded 50,000 square miles.[147]

To harness the Yellow River—described in full flow by an American reporter as "a giant angry mudslide"—Chinese engineers have recently straddled it with the massive Xiaolangdi Dam, 4,320 feet long and 154 feet high.[148] Its innards contain sixteen gigantic steel-lined tunnels designed to flush the impossibly heavy silt-load through the dam, while channeling the clearest water through giant turbines. Because of the huge sediment loads, Xiaolangdi's useful life won't be more than thirty years; after that the loess will have consumed its reservoir and begun to climb up the walls of the dam itself. Chinese engineers hope by then to have built a number of other dams along the Yellow.[149]

But most of China's attention is focused on the Yangtze—the Changjiang, in Mandarin—which spills off the roof of the world in Tibet into a maze of mountains and tributaries running all the way to Shanghai and the East China Sea. Changjiang means "Long River"—at 3,450 miles, only the Nile and the Amazon travel farther. It is as unpre-dictable as it is long. Records dating back to the Han dynasty indicate that the Yangtze can be counted on to flood severely every ten years along its upper stretches and experience major floods every five years or so in its central basin, which contains China's most agriculturally productive—and its most intensely crowded—land.[150] If it were a nation, the Yangtze Basin would be the second most heavily populated in the world. One in every twelve people on the planet lives along this river's banks—more than 500 million people crowd onto its watershed.

The most celebrated—and notorious—of all China's planned projects is the Three Gorges Dam on the Yangtze, which, when completed in 2003, will be the world's largest concrete structure and its most powerful. Not since the Chinese built the thousand-mile Grand Canal have they moved so much earth.

The project began under the Nationalist regime in consultation with the U.S. Bureau of Reclamation. In 1944, John L. Savage, chief engineer of the Hoover Dam, drew up the first plans, calling for multiple dam sites in the Gorges. The Communists proved less patient. "If so much effort is needed to construct dams on the tributaries," wrote Chairman Mao, "and still it is not enough to achieve the objective of flood control, why don't we just concentrate on the Three Gorges and stop the floods there?" It fell to Premier Li Peng, himself a hydraulic engineer, to launch the project by throwing the first shovel of dirt in Xiling Gorge near Yichang in 1994.

The impulse to build Three Gorges derives directly from the stark facts of Chinese life: 1.3 billion people, a fifth of the world's population, live here, on an eighth of its land. Its cities are bloated, its air is fouled by burning coal, and its earth is poisoned by showers of acid rain. Chinese officials have made the same encouraging noises about the project that have always been made about dams: an end to floods, enough electricity for all of south and central China, and an enormous boost to the economy of the whole region.[151]

But the damage Three Gorges seems certain to bring will be staggering. The forested crags that rise up to 4,000 feet above the river—celebrated in centuries of paintings and attracting 17 million awestruck visitors every year—will be lopped off at their knees, losing 500 feet to the rising reservoir. The dam will do away with rapids and whirlpools, turning the river into a 400-mile-long lake that will obliterate 140 towns, 4,000 villages, 32,000 acres of precious farmland, and the irreplaceable architectural record of thousands of years of Chinese history. And it will displace no fewer than 3 million people, who will have to build new lives for themselves somewhere else.

Critics also point out that the big dam is situated dangerously in a seismically unstable location prone to landslides. Reservoir-induced tremors—set off by the mountainous weight of water or seepage into the foundation caused by pressure—don't often rise above a 3.0 on the Richter scale, but there has already been at least one frightening exception in China: in the 1960s, the filling of the Xinfengiang Dam triggered 250,000 small earthquakes and one major one, measuring 6.1. Even without the threat of an earthquake, China's dam safety record in recent years has not been encouraging: more than 3,000 dams built in the Mao era have broken because of shoddy materials or poor design. In 1975, two dams in central Henan Province gave way under 26 hours of pounding rain, killing more than a quarter of a million people.[152]

The problems caused by silt will be nearly as formidable at the Three Gorges as they are at Xiaolangdi. Still more alarming is the fact that 265 billion gallons of raw sewage are dumped into the Yangtze every year, along with runoff from thousand-year-old landfills and factory waste from Chongqing. This noxious blend will build up behind the dam, creating a malevolent cesspool, a source of dangerous disease.

Undeterred, China keeps on building out of desperation—and unexamined hubris. The brand-new Gezhouba Dam, which already sits astride the Yangtze just 25 miles downstream from Three Gorges, has the largest spillway capacity of any dam in the world. Twelve more big dams are planned on the upper Yangtze, as well as twelve oversize hydroplants on the Han River, a Yangtze tributary that joins it downstream of the Three Gorges. Finally, in the most ambitious scheme of all, a river diversion project costing tens of billions of dollars will heave three major canals thousands of miles across the country, carrying water from the Yangtze River north to the Yellow, from the mountains of Tibet to the cities of Beijing and Tianjin. The Chinese government means to bring its rivers under control at any costs.

Will the dams stop flooding? An American team of experts surveying the Yangtze Basin in the 1980s didn't think so. The Tennessee Valley Authority was able to control flooding by building a series of dams along 650 miles of the Tennessee River. But the Yangtze is over five times as long, its flows greater in multiples, and it receives heavy rains during monsoons, burdening already swollen rivers moving down from the mountains. And while there are scores of tributaries on the Tennessee, seven hundred tributaries pour into the Yangtze River, many of them downstream of the Three Gorges Dam. In flood season, large tributaries like the Min or the Yalong enter the Yangtze with the force of tidal waves. During the flood of 1954, the Yangtze discharged twice the capacity of the Three Gorges reservoir.[153]

The life-and-death struggle against water has been so important to the Chinese that one of their most revered figures is an engineer, Yü the Great. Yü was said to have carved the twisting path the Yangtze follows through the limestone gorges of Central China, and then to have been made emperor by a grateful people. "But for Yü," the Chinese say, "we should all have been fishes." Five thousand years after the warrior engineer Yü was said to have drained the land, almost half of all the big dams in the world have been built in China, and still the country digs. No one knows if Yü the Great was a real man or merely a legend, but a statue of the warrior engineer, gripping a simple digging tool, still stands on a cliff near Zhenzhou, gazing out over the treacherous Yellow River. He is an eloquent reminder of man's seemingly endless struggle to find new ways to coexist peacefully with the world's great rivers, a search that will only intensify as more and more people crowd onto the land. In Bangladesh,

China, America's heartland, and countless other places, that quest will require intelligence as well as energy, imagination as well as ambition. China's proposed solutions seem especially troubling; if we are not careful, in her uncertain future there could be whispers of our own.

Chapter 7

THE WARS

*What we call man's power over nature turns out to be a power
exercised by some men over other men with nature as its
instrument.*

C. S. LEWIS, *THE ABOLITION OF MAN,* 1947

During a visit I made to southeastern Anatolia in the fall of 1989,
Syrian MIG fighters shot a Turkish survey plane out of Turkish skies,
killing five people. Everyone seemed to believe that the incident had
been meant as a warning over the Ataturk Dam, which was being built
on the Euphrates River. If it was a warning, the Turks were unmoved:
they announced that Syria and Iraq had better start storing water since
they were about to stop the flow of the Euphrates to fill Ataturk's vast
reservoir. Turkey insisted publicly that its neighbors would "understand"
this necessity. Predictably, their neighbors' understanding has not been
profound.

Caught up in the painful politics of nations, water—and the question
of who controls it—has already been a factor in igniting at least one full-
scale conflict—the Six Day War of 1967, when the Arab League,
angered at Israel's construction of its National Water Carrier, which had
appropriated much of the water of Jordan River for use in Israel, began
to dig canals to divert two Jordan tributaries, the Hasbanin and Wazzani

Springs. Israelis immediately shelled and destroyed both projects. The attacks by Syria, Egypt, and Jordan that eventually followed had many causes, but water remained a priority for both sides. Before it ended, Israel had blown up a dam Syria had been constructing on the Yarmouk River and annexed the Golan Heights; it took the West Bank from Jordan along with one-third of that kingdom's most fertile land; and it seized from Egypt both Gaza and the Sinai Peninsula. All of it, except the Sinai, secured precious water for Israel.

Even though many people are without adequate water, hostilities haven't generally become armed conflicts, sometimes if only because military strength is concentrated heavily in the hands of a few. However, if, as *The Economist* says, "wars are fought over much stupider things," it seems plausible that in nations under stress, water can be the push that carries vexatious neighbors over the brink.

A river is without nationality, even though its name changes as it crosses from one country to another. "If you left it to the rivers," Willson Whitman said, "nations wouldn't fight at all." But it is no longer left to the rivers. Almost half the earth's land lies in river basins shared by at least two nations, and 80 percent of the world's available fresh water flows through international river basins. Some 261 large rivers pour from one nation into another. Most countries share water more or less amicably but in some of the world's most water-short regions, particularly the Middle East and parts of Asia, the strains are beginning to show. The Middle East alone is home to five percent of the world's population and only 1 percent of its renewable water resources.

THE TIGRIS AND EUPHRATES RIVERS, which flow southward from the wild hills of Central Turkey to the Arabian Sea, once watered the lush region known as the Fertile Crescent. Now, a bitter quarrel over ownership of their water, fueled by steadily escalating demands upon them, threatens two of the three powerful nations through which they run—and in the process poses a grave danger to the peace of the Middle

East. The Southeast Anatolia Project—an immense engineering scheme that will place twenty-two very large dams and reservoirs on the Tigris and Euphrates rivers at their source in Turkey—will catch the rivers in a viselike grip and threaten the survival of Turkey's downstream neighbors. The Euphrates is Syria's fundamental water source. Iraq, through which both rivers run, must make do with what Turkey or Syria have not already siphoned off. Since there is no formal three-way agreement on sharing these waters, Turkey's downstream neighbors are understandably concerned. "We were also nervous when we were short of oil," Turkish Minister of State Kamran Inan responded when I asked him in 1989 about anxiety in his neighborhood. "Water here is much more rare than oil, and it will be more and more so." Ferruh Anik, head of the Turkish government agency that is building the project, is still more pointed: "The water doesn't flow for free. The Arabs have to understand that."

At the core of the tensions in the Jezirah—the Arabic name for the land between the rivers—is the Ataturk Dam, a mass of earth and rock, which blocks the Euphrates on a mighty scale. I visited Turkey twice during the late 1980s while the big dam was going up, and was staggered by its scale. Night and day, the construction site throbbed with the roar of a $300 million fleet of graders, bulldozers, cranes, and 1,775 trucks. Busloads of workers, 8,000 a shift, came and went, transforming 85 million cubic meters of rock and earth into a man-made mesa more than a mile long. I heard a visitor gasp at the sight: "These dams are our pyramids." Mighty Ataturk Dam, a $3 billion project, ninth largest in the world in bulk, is a monument to Turkish initiative and a source of enormous Turkish pride.

After circumnavigating Ataturk's hilly landscape in a small car, a friend and I drove off to explore two tunnels under construction upstream, each wider than a subway tunnel and longer than Manhattan. I'll never forget having to back out of one of them as a tunneling machine the size of a small house moved inexorably toward us from the other end. These irrigation tunnels, said to be the world's longest, would soon carry 328 cubic meters of water each second out of Ataturk's reservoir to irrigate 1.5 million acres around the Harran Plain. Behind Ataturk, the reservoir would soon put everything within 315 square miles under water. The river valley was eerily quiet, the brown-gray hills of the upper Jezirah cloudlike and still.[154] Scattered across the flat land along the Euphrates were segments of Roman aqueducts, reminders that great migrations of peoples swept back and forth across this ground over centuries. Ancient empires had collided here like continental plates.

This project will be accompanied by irretrievable losses. Scores, perhaps hundreds of ancient sites will be flooded. One of the most important is the former Seljuk capital of Hasankyef, doomed by the projected Ilisu Dam on the Tigris. Archaeologist Olus Arik is mounting a last-ditch effort to excavate the city and the ruins that surround it before they disappear forever beneath the waters. "We must learn what this city was," he insists. "Ilisu Dam is only for electricity. I am told it will silt up within seventy years. For the sake of a few million kilowatts we will lose a center of two thousand years of culture."

But another scholar, Nimet Osguç—who did similar work at Samsat, an earlier site now inundated by the Ataturk reservoir—probably speaks for most citizens of southeastern Anatolia, for whom such losses are outweighed by the project's promised benefits. Although it broke her heart to see Samsat vanish, she says, "No one can say, 'Don't make this.' Not even though it put my wonderful city under the water. This area will reach prosperity."

The Jezirah once knew prosperity well. "This territory," wrote Herodotus, "is, of all that we know, the best by far for ripening grain." But as we drove across the empty, parched landscape, it was clear that things had long since changed, that there was a desperate need for water to replenish the exhausted land and stop the wholesale exodus of failed farmers and their offspring to the country's already overcrowded cities. William Willcocks, who devised an early-twentieth-century scheme for irrigating the Jezirah with water from both rivers, was sure it could be brought back to life. "She has always risen with an energy and thoroughness rivaling the very completeness and suddenness of her fall," he wrote. "She has never failed to respond to those who have striven to raise her." Her encouraging response may be clearly seen in the region around the city of Sanli-Urfa. Euphrates water, brought in through Ataturk's giant tunnels, has restored this region, believed by its residents to have been the site of the Garden of Eden, to something like its former glory. Green fields filled with cotton, corn, beans, sorghum, alfalfa, and several varieties of melons now flourish in a landscape that just a few years ago had been brown and parched.

In Sanli-Urfa, at the end of my visit to the Southeast Anatolia Project, I stopped to see an old friend, Ali Balaban, project advisor to the Turkish government and chairman of the Agricultural Engineering Department at the University of Ankara. The project was everything Turkish politicians promised it would be—and more, he told me. With all the dams and irrigation systems in place, it would water and electrify an area larger than Belgium, Holland, and Luxembourg combined. When we spoke, the Southeast still accounted for just 4 percent of Turkey's irrigated land; eventually that figure should rise to 64 percent. But the heavily mustached Balaban is most fiercely proud of the additional benefits that will be brought to a broad segment of the population—roads, railways, container terminals, telecommunications, clean drinking water, sewage and wastewater treatment, industrialization, and the modernization of agriculture. "We will completely transform this land physically and socially in the next thirty years," Balaban told me blissfully. "Now, we get one crop in two years. But we have sun and if we add water and technology to such rich earth, we can grow two or three crops every year."

Like every great engineering scheme, the Southeast Anatolia Project is driven by politicians, and even in 1989 was swallowing up roughly $2.5 million a day. I wondered aloud whether the opposition wouldn't exploit that fact. Balaban wasn't worried. "In a democratic country like Turkey, power can certainly change hands," he said. "But this is a national project. The work will go on."

Minister of State Kamran Inan agreed. An elegant, gray-haired power in Turkish politics, he had been in charge when Turkey first flung herself with full force into the reordering of land and water along the Tigris and Euphrates, and he had not lost his enthusiasm for the scheme. "In the coming decades in the Middle East, the most important resource will be water," he told me, "and we are the richest possessors of the resource in the region. These rivers have been here for millions of years. We want to put them to use to benefit the children of this country."

When Ataturk went online in July 1992, a triumphant President Turgut Ozal stood on its ramparts, fireworks exploding overhead, and exclaimed, "The twenty-first century will belong to Turkey." That kind of talk does not please Turkey's neighbors Syria and Iraq—and for good reason. The Southeast Anatolia Project may deprive Syria of half its Euphrates water and will pollute the rest with fertilizers, pesticides, and salts. It will also cut off nearly 90 percent of the waters that have traditionally flowed into Iraq, where rivers are already strained, aquifers are overpumped, and cities are parched. The Syrians, who have already begun to riot over water and food, have, in turn, begun their own engineering projects, which may eventually reduce Iraq's waters and fill them with still more pollutants as well.

Turkish politicians are unsympathetic. "This is a matter of sovereignty," former Turkish Prime Minister Suleyman Demirel has said. "Water resources are Turkey's and oil is theirs. Since we don't tell them, 'Look, we have a right to half your oil,' they cannot lay claim to what's ours. These crossborder rivers are ours to the very point they cross the border."

In the early 1970s, Syria and Iraq themselves almost went to war over the waters of the Euphrates. When Syria built a dam at Tabaq—denying a quarter of the river's flow to Iraq while its reservoir filled—and further refused to negotiate over water rights, troops massed on the borders, and only intervention by Saudi Arabia and the Soviet Union kept them from attacking each other. As tense as relations between the two countries have become, they remain united in their opposition to Turkey and the support both countries have shown for Kurdish separatists in that country is in part retaliation for Turkey's water policy. In 1988, Syrian and Turkish troops faced off in what a Turkish general called "an ongoing state of undeclared war." Four years later, Iraq officially accused Turkey of "pirating" the Euphrates and denounced the Ataturk Dam as a "violation of international law."

There are some in the region who believe there is enough water for everyone to share and it is merely stubbornness that prevents an amicable solution. "If it were left to technical people," a Syrian water official has said, "within three months we could reach an agreement to guarantee everyone's needs." Tevfik Okyyuz, a former Turkish minister of foreign affairs, argues that the Southeast Anatolia Project itself will ultimately benefit everyone. "The plan dictates that all the dams be constructed in Turkey for climatic reasons," he says. Development should be pursued according to what the land can sustain—if, for example, experts find that Iraq is more suited to the cultivation of cotton than Syria or Turkey, then Iraq should grow the cotton. Turkey, which can better grow beans, would sell its beans to Iraq and buy Iraq's cotton in return, and so on.

Even if irrigation schemes are better adapted to Turkey's soil and climate, the interconnected nature of the watersheds means that only by sharing technical data can the three "cousins" ensure water security. Syrian Foreign Minister Farouq al-Shara has insisted that nothing can be accomplished without a formal three-way agreement. "What concerns us is not the quantity of water that passes through the Euphrates, but the lack of a comprehensive treaty on water," he said after a meeting in Istanbul. "Sharing the water resources of international rivers is of importance to any country. We want international laws and norms to be applied to the three nations."

Any talk of sharing is fanciful without political reconciliation.[155] These neighbors have complicated, sometimes hostile histories, making it difficult to resolve even issues critical to their survival. Of the three, Turkey has the most water and believes in its right to use it, no matter how high the political price.[156] "We no longer count on the uniquely strategic importance of our location," said Kamran Inan. "We are now willing to be more of an economic and technological pioneer in this region by using our raw materials. . . . This," he adds, "will make Turkey a candidate for a position of power in the Eastern Hemisphere."

Much of the Middle East needs Turkey's resources, but so far only Israel has come to terms with this fact. At a meeting in 1997, representatives from seventeen Arab states and Palestine gathered in Damascus for a conference on water. Neither Turkey nor Israel was invited. Israel has made an agreement with Turkey to buy substantial quantities of water to be towed across the Mediterranean in Medusa bags. And America's Bechtel Corporation has had discussions with Israel over the feasibility of building water and oil pipelines to Turkey.

But Turkey continues to play a dangerous game by building dams that will deny its closest neighbors access to the waters of the rivers. "I'm very excited for the Turks," John Kolars, an engineer who has worked extensively in this region, told me in a phone interview. "But if you see your brother doing something foolish, you must tell him. They have to take their downstream neighbors into account."

Turkey's obstinacy can't be separated from larger security concerns in the region. "Water affects all of the peace process," hydropolitical analyst Dr. Joyce Starr said to me a few years ago in Washington. "Because Syria is so affected by its relations with Turkey, it can't possibly come to a satisfactory water relationship with Jordan and Israel." Starr, a forceful and attractive woman who served as an advisor to President Jimmy Carter during the Camp David talks and ran the Secretariat for the Reconstruction of Lebanon under President Ronald Reagan, persuaded Egypt's Boutros Boutros-Ghali to coconvene the first African Water Summit in 1990 in Cairo. The forty-two nations attending developed the Cairo Water Declaration, which asserted that through cooperation, Africa could stretch land and water resources to support its growing population. From that effort, Starr went on to probe ways of solving hydropolitical problems around the globe. "In the best of all possible worlds," she told me, "there would be a regional authority, which would channel resources to institutions in the regions that are already engaged in the work, but need to be strengthened, institutions like those in Israel and Egypt doing good work on arid land agriculture that could play more regional roles. Statesmen want to show big gains. Especially around election times—maybe a desalination project here or a wastewater-treatment project there. But where is your environmental impact statement for the desalination plant? Is this the best approach? The first step is parallel approaches—by the different parties. Now this is drudgery of the first order to start this dialogue. You need to look at the many options. A creative approach to water cooperation could forge a new path to peace."

Whatever arrangements nations are able to make with each other, no one doubts anymore that lasting peace is possible in the Middle East only if water is taken into account. "It is vital to the economic and political survivability of the region," says Starr. "Middle East hatred is bountiful but Middle East water is at the point of no return."

Although history still reveals a few full-scale wars fought over water, thirst is an important cause of bloodshed within national borders.[157] A recent research project, funded by the American Society for the Advancement of Science and the University of Toronto, studying the links between environmental scarcity and violence, produced a chilling report. It concludes: "Severe environmental scarcities often contribute to major civil violence. Poor countries are more vulnerable to this violence, because large fractions of their populations depend for their day-to-day livelihoods on local renewable resources, such as cropland, forests, lakes, and streams. . . . Moreover, poor countries are often unable to adapt effectively to environmental scarcity because their states are weak, markets inefficient and corrupt, and human capital inadequate. . . . The result is often chronic and diffuse subnational violence that is exceedingly difficult to control using conventional means, that undermines development, and that sometimes jeopardizes the security of neighboring countries."[158]

The report, taking Pakistan as a case study, went on to detail the factors that lead to violence. Pakistani women give birth to an average of 6.6 babies per woman, a population growth which cannot be supported by its exhausted soil. Its farmers suffer further from a host of water-related problems: scarcity, salt loads, waterlogging, pollution, and bad management. Crop productivity has steadily diminished as a fast-growing population attempts to live off too little land. The rural poor, badly educated and politically powerless, progressively make their way to the cities, where the state is unable to take care of them. Criminal gangs take up the slack, controlling housing and food, and "tanker mafias" control the water supply. The result of such progressive misery has been persistent violence in Islamabad, Rawalpindi, and Hyderabad. Karachi—Pakistan's largest city—is battered from frequent rioting and brutality. "Environmental scarcity can contribute to diffuse, persistent, subnational violence, such as ethnic clashes and insurgencies," writes one of the report's editors, Thomas Homer-Dixon. "This [internal] violence is not as conspicuous or dramatic as interstate resource wars but it may have serious repercussions for the security interests of both the developed and developing worlds."

OF ALL THE WORLD'S water-deprived regions, none encompasses more conflicts, ongoing and potential, than the lands along the Nile. It is the world's longest river and comes barreling out of the African High-lands, half a continent, 4,238 miles away from its Mediterranean delta. No other river on earth flows through such diversity. The river is not one but many: the Victoria Nile, the Albert and the Sobat, the Bahr el Ghazal, the White Nile, and the Blue Nile. The lands of the Nile's watershed are also many and varied—the slopes of the Mfumbiro volca-noes in Zaire, Uganda's Mountains of the Moon, and the sumptuous Masai Mara and Serengeti in Kenya and Tanzania. Hundreds of different languages are spoken in this basin, but its distinctive societies are alike in their growing reliance on the waters of the Nile. What these people build on the river will write much of the story of their future in the next century.

Two-thirds of the African population is threatened by water scarcity and famine. There are more international river basins in Africa than in any other continent, fifty-seven in all. But nowhere in Africa are the tensions of drought, inadequate water supply, and water politics felt more acutely than along the Nile River. Whenever engineers turn up along the Nile, the tremors are felt all along its reaches, nowhere more so than where the river ends in Egypt. Here all life is the river's gift—it couldn't be clearer. Move away from the river and green turns to brown, movement ceases, life stops. But while Egypt and the Nile have long been said to be as one, ten African countries share Nile waters. Each wants a piece of the river.

Not long after signing a peace accord with Israel, Egyptian President Anwar Sadat said that only water could make Egypt go to war again. In 1990, when Israeli engineers were discovered to be investigating the feasibility of three dams in Ethiopia, Egypt's Deputy Prime Minister for Foreign Affairs Boutros Boutros-Ghali warned that the construction of a dam on the Blue Nile would be understood as an act of war. Today, new dams have been proposed in the Sudan, the Congo, Tanzania, Uganda, and Ethiopia.

American scholar Robert O. Collins, who has explored the Nile from its most distant springs to its wide delta, believes there is just one truly pertinent issue along the Nile—the question of who possesses water rights. "There are two views, the first being that it should be used for those that require it the most," says Collins, an imposing man, who now lives in California. "Most Angelenos agree with that position. For the Egyptians, this is the number-one principle—it goes back as far as Dynastic Egypt—that they have a right to the water: historic, acquired, and established.

"The opposing view is that it should be developed as a basin," Collins continues. "That all the riparian peoples in the basin have a right to the water even if they don't yet use it. This view is usually described as equitable utilization, or fair use, and it means they all have a right to a chunk of it. Which is it going to be? Do you develop it as a basin or do some people have greater, prior historic needs and consequently a greater right to the waters?"

All of the countries of the Nile—Ethiopia, Eritrea, Burundi, Rwanda, Zaire, Kenya, Tanzania, Uganda, the Sudan, and Egypt—are poor and growing, seven of them among the world's least developed nations. They are all primarily agricultural, dependent on Nile water, and afflicted with soil erosion, drought, and land degradation. Even now, most of their trade is outside the region, not with each other. They all receive international aid, bear large debt loads, and apart from Egypt none is wealthy enough to build large water-control projects without foreign financial and technical help.

And yet Egypt's nine upstream neighbors are deeply suspicious of the thirsty country at the end of the line. In the past, Egypt has offered money and technical expertise to increase Nile flows, assisting with works like the Inga Dam in Uganda, on the condition that they have a hand in running it. One of Egypt's ideas for regional Nile management is to barter electricity from the Aswan and Owen Falls dams in exchange for guaranteed water flows. Yet over many years and after many meetings, only Egypt and the Sudan have signed a Nile treaty, and that was more than forty years ago.[159] At the same time, Egypt has chosen to divert water hundreds of miles to new irrigation projects in its deserts, even though other attempts at desert settlements have failed and the water losses to evaporation will be astronomical. In 1997, President Hosni Mubarak opened four large tunnels under the Suez Canal to carry Nile water into the Sinai Peninsula and initiated the construction of a pumping station behind the High Dam at Aswan to send water into the Western Desert, calling it a "National Project."[160] Egypt's neighbors consider it a national fantasy.

When I asked Collins about the irrigation settlements, he expressed apprehension. "They are dropping a lot of water in a very large desert," he said. "Where is that water going to come from? Egypt says that it is conserving water; that the United States is paying millions of dollars to reconstruct the Cairo water system; that they are redoing their irrigation canals to make them less wasteful; and that they are using drip irrigation in the desert. But I am suspicious." Collins concedes that Egypt is making efforts at conservation, but does not think that even the Egyptians seriously believe they can succeed without still greater flows. "What they believe," he says, "is that they have got to find additional water."

Egypt has historically placed its trust in cement and stone. In the twentieth century, the British sought to secure themselves in Africa through engineering on the Nile; later, the Egyptians took their future into their own hands. At the heart of both efforts were dams at Aswan. So it is not surprising that Egypt has initiated a collaboration with the Sudan based on concrete—the formidable Jonglei diversion project meant to sprawl across one of the largest swamps in the world and one of its most inaccessible places, the Sudd.

The Nile slows almost to a stop in the 3,000 square miles of the Sudd, its wild, flat lands teeming with crocodiles, birds, half a million wild hoofed animals, and a quarter of a million people.[161] The proposed Jonglei Canal, a cooperative attempt between the Sudan and Egypt, would straighten the Nile's disorderly course through the Sudd, shortening the water's distance by about two hundred miles. Because about half of the flow of the White Nile is lost to evaporation in these swamps, the canal would bring an extra 20 million cubic meters of water to Aswan every day.[162]

But the Muslim north of the Sudan has long been engaged in a civil war with the southern tribes. In 1976, when a French consortium began digging the canal, there were riots in Juba in the southern Sudan, and the Sudan People's Liberation Army attacked the canal. In 1984, with 166 kilometers dug and 60 to go, it was stopped entirely when, in an attack on the base camp, one engineer was killed and the others routed. The Egyptian-Sudanese working relationship also ground to a halt.

"The people of the canal zone have the worst of all possible worlds," says Robert Collins, "a land ravaged by war and an empty canal providing neither water nor benefits." But he does not believe that the canal is a dead issue. "The directors of the French Canal Company put it to me very simply," he says. " 'We are French. We are canal builders. We were at Suez. We built at Panama. We will build this canal. Money is not the issue.' But they cannot get on with the job so long as the Sudan is at war with itself. All of this is flim-flam unless you stop the fighting.[163]

"The Sudanese and the Egyptians don't like one another," says Collins. "The Egyptians are upset about a Sudanese wish to build a dam at the fourth cataract, but they are cautious about being heavy-handed with their Sudanese brothers. Egypt has exercised restraint both in this relationship and with Ethiopia. There are finally two players that matter here, Ethiopia and Egypt."

Ethiopia, with 84 percent of the Nile water gushing out of its green, rocky gorges, is really the king of the river. William Willcocks once wished for a great dam at Lake Tana in Ethiopia—then Abyssinia—capable of holding an inexhaustible flow of water in a climate where it would not be lost to evaporation. "But," he concluded, "it might not be convenient on political grounds to put one of the great public works of Egypt at the absolute mercy of the Abyssinian Emperor."

Others have had similar thoughts. In 1958, the same year in which the Soviets offered to finance the Aswan High Dam, the U.S. Bureau of Reclamation turned up in Ethiopia and proposed four dams on the Blue Nile. In the 1970s, the USSR conducted feasibility studies around Lake Tana, and Israeli engineers have since investigated water control on the Blue Nile as well. Dams on the Blue Nile, where evaporation would be minimal and water releases could be timed to coincide with dry seasons in the lower basin, could benefit downstream countries considerably. "If, in a Utopian world where everybody behaved well toward their neighbors," says Collins, "you constructed appropriate dams in the Blue Nile

Basin, you could acquire more water. Now we know that's years away and will require many negotiations."

Ethiopia, devastated by malnutrition and with a birth rate of 6.9 babies per woman, higher even than Pakistan's, desperately needs irrigation development and, even more than some of its downstream neighbors, electricity. Although Ethiopia and Eritrea have seen some good rains in recent years and have even exported grain surpluses, these two countries are among the poorest of the poor. That they have continued to fight with each other in the face of dire poverty—a war that *The Economist* likens to "Two bald men fighting over a comb"—has kept them from building large engineering works. "The Ethiopians possess the water," Collins remarked. "But the Egyptians know that nothing will be done immediately in Ethiopia, because of its instability, its poverty, and because the international agencies and governments which might supply the money will not do so unless there is an agreement."

Although so far Egypt has not gone to war with Ethiopia to protect its water supply, it has repeatedly stated its willingness to do so. In 1980, when Ethiopia complained to the Organization of African Unity that Egypt was misusing its water rights by diverting water to the Sinai, Anwar Sadat made things crystal clear. "If Ethiopia takes any action to block our right to Nile water," he said, "there will be no alternative but for us to use force. Tampering with a nation's rights to water is tampering with its life, and a decision to go to war over this issue is indisputable in the international community."

In the meantime, the population of the region is steadily rising. There are currently some 250 million people in the Nile Basin; in twenty-five years that number is expected to double. Figures like that have motivated the World Bank to try to find a mechanism by which international financing for multinational water projects can be arranged and the needs of all the Nile nations can be met. "The World Bank is proud of what was accomplished in the Indus Basin," Collins concludes. "Reconciliation is not easy in such a highly charged situation, but the bank has a lot of leverage. Perhaps the weight of history lies too heavy in the silt of the Nile Valley, but man will always need water. And in the end, this may drive him to the river to drink with his traditional enemies."

IF THE NILE NATIONS do finally sit down together to work on their differences, they will have precious few universally recognized precedents to build upon.

As early as the second millennium B.C., the Babylonian Code of Hammurabi provided punishment for those who stole water or neglected irrigation systems. In some legal codes, such as Roman-Dutch law, water belonged to the state. English common law tied water rights to land rights, granting those who lived along a river access to the water but stipulating that no riparian landowner could diminish either the quality or the quantity of water for downstream users.[164]

In areas subject to drought, law and custom tended to be more flexible. The Tswana, cattle-herding people in southern Africa, for example, accede water rights to that tribe or group that has been "allocated" the pasture land. Herders passing through the pasture land are given the right to take water for themselves and their cattle but must obtain permission if they wish to linger—a remarkably sensible ordering of water use, developing out of respect for the need for water to sustain life.

"In the wise old Eastern world, water has duties attached to it as well as rights," remarked William Willcocks. Water was of paramount importance to early societies in the Middle East, and sharing it was central to the survival of its people. Given the dry nature of those lands, it is no surprise that *Shari'a*, which today refers to Islamic law in general, originally meant the law of water. *Shari'a* identified water as God's gift, which no one may deny to another.

As famously complex as water rights have always been, concepts of ownership have become even more complicated as man has learned to control it. Although Islamic water law regulated irrigation ditches and wells, spacing them so that no one's water supply would be infringed upon, it's difficult to see how these rules could be made to apply to today's intricate waterworks, especially when they touch upon the waters of more than one country:[165] "The customary rules are diffuse," writes Dr. Hasan Chalabi of the University of Lebanon. "They are the obligations of good will, of good sense, and dependent in large measure on the cooperation and the good deeds of riverside dwelling states."

Good will, good sense, and good deeds are rare in international relations, and it is no surprise that there are as yet no real, binding international legal mechanisms to settle water disputes, any more than there are foolproof mechanisms to stop nations from fighting with each other. Still, there have been a number of attempts to codify the basic tenets of water-sharing across borders. Perhaps the two most important were the Helsinki Rules, developed by the International Law Association in 1966, and the thirty-two articles issued by the International Law Commission of the United Nations in 1991. The documents differed in their details but both urged that international disagreements over water should be settled on the same basic principles: shared rivers must be seen as international watercourses, not the exclusive property of any one country; all riparian nations have a right to water (the principle of "equitable utilization"); no nation's use of shared water may be allowed to damage the well-being of its neighbors; and finally each must provide accurate hydrological information to the others. These principles have been applied successfully to several disputes, including the question of water rights in the Komati River basin, which twists out of South Africa into Swaziland and Mozambique then back again. They were also used in a recent judgment by the International Court of Justice concerning a dispute over the Gabcikovo Dam on the Danube River between Slovakia and Hungary. The court in this case invoked equitable utilization, joint management, environmental standards, and the United Nations Convention on international water law.

International water law still lacks fact-finding machinery and means of enforcement. Palestinian water expert Sharif Elmusa, who served as advisor to his delegation during talks that led to the Oslo Accord and the

1995 Taba Agreement in Washington, which achieved agreement on the all-important principles of equitable utilization of water rights and joint management, believes that the mere existence of agreed-upon principles has at least limited power of its own. "It's an ongoing battle, but without international law how can you make your demands?" he asks. "First of all, an upstream country like Turkey may *say* that the Euphrates is a Turkish river, not an international one, but international water law declares it to be international since it is located in three countries. To say this is an international river, rather than 'my river,' is a very important first step. Then, too, international water law recognizes no right of prior use or absolute sovereignty: Egypt or Israel may claim that existing uses should dictate how waters are divided in the future, and Turkey may say, 'We will take what we want from our territory.' But nations don't like to appear to be going against widely accepted policy. They may say, 'We are adhering, but this is our interpretation.' No one will say, 'We are not adhering to international law.' "

The same principles recently helped calm a serious struggle in the Iberian Peninsula. During the mid-nineties, the worst drought in a century gripped the region. Roughly half of Portugal's water flows westward from Spain, and when the Spanish government, beleaguered by a series of dry years, announced ambitious plans for a series of new dams that would have held on to much of that water—and dangerously reduced her neighbor's supply—Portugal called for talks. International water law made it all possible. "We had an important framework for negotiations," recalled Pedro da Cunha Serra, president of the Portuguese Institute of Water. "The European Union has issued directives on water. So have two United Nations Conventions on Trans-Boundary Water Courses. And the fact that both countries have to respect the framework of legal obligations made it much easier for us to forge an agreement." On November 30, 1998, after five years of drought and discussion, Spanish and Portuguese ministers signed a bilateral treaty that allowed Spain to construct new dams and at the same time guaranteed

Portugal sufficient water from the dammed rivers to meet its needs. All the outstanding issues between the two nations were not solved, but each country is now free to fix its attention on solving its own internal battles against persistent drought, antiquated water systems, and agricultural overload. "A treaty?" said da Cunha Serra. "It would not be possible to resolve our water problems without it."

International water law has held out at least one hint of hope even for the mutually antagonistic nations of the Jordan Basin—Israel, Syria, Lebanon, and the kingdom of Jordan itself. Here, there is no unclaimed water, no untapped stream or virgin spring or undiscovered groundwater. Rivers run shallow, their scant flow laced with salts. Aquifers are overdrawn and the rains that could recharge them are unpredictable at best. And the population is growing faster than that of any region other than Africa.

Because of overpumping, Israel's main sources of water are being depleted faster than they can be replenished.[166] At the mercy of Israel on the Jordan, and Syria on the Yarmouk, the people of Jordan have less water than any place in the Middle East. Many towns in Jordan receive water only twice a week, yet their water needs will double in the next two decades.[167] In areas of the Palestinian Authority, access to aquifers is severely controlled and curtailed by Israel. Many villages, without any water supply, are forced to collect occasional rainwater in cisterns and barrels or to wait for deliveries from trucks. In recent years, the lack of water has shriveled Arab farms in the West Bank. Now less than 4 percent of these farms are irrigated. Wells here are drying up because Israeli settlers' deep wells and powerful pumps drain the water while Palestinians are forbidden to dig new wells or even deepen old ones.[168] In the words of one Palestinian, "No one can accept that he does not have water to drink and his neighbor has a swimming pool."[169]

When it comes to water, no one gains in the Jordan Basin without someone else losing. There has already been at least one bout of armed conflict in this region caused in part by the struggle over water. Yet even here, thanks to the principles of international water law, Israel and Jordan, at least, have moved a little way toward fairer allocation, which might one day help point the way to genuine peace.

The late King Hussein of Jordan understood this as early as 1964, even as Jordan and its Arab neighbors sought to divert waters away from Israel for their own use, one of the incidents that led to the Six Day War. "We were talking about two peoples who were destined to live together in a very small region," he remembered saying at the time, "and who had to figure out how to resolve our common problems. . . . Every aspect of our lives was interrelated and interlinked in some way or another. And to simply ignore that was something I could not understand. One had to do something, one had to explore what was possible."[170] That exploration took thirty years, but the 1994 treaty that finally ended the state of war between Israel and Jordan and established full diplomatic relations between them also provided each country with a specified amount of water every year.[171]

Few things go smoothly in the Middle East, and the 1994 agreement has proved no exception: in 1999, when rains in the region were even more sparse than usual, Israel's water commissioner, Meir Ben Meir, announced that Israel would have to cut its promised supply of water to Jordan but offered to share the deficit equally; Jordan replied firmly that it would accept no cuts. Still, the 1994 treaty represented a cautious first step toward the kind of joint river-basin management that the region so badly needs. "Allocation is the single most important issue, not just in the Jordan Basin but all over the Middle East," says Sharif Elmusa. "The question in joint international basins is how much everyone will get. Only the peace process can address this."

I asked Elmusa about something he had written in his 1997 hydropolitical study of the region, *Water Conflict*: "Scarcity and dependence on common resources are a sure recipe for conflict." Did he mean that all-out war over water was inevitable in the Jordan Basin? "Conflict doesn't always mean military conflict," he answered, measuring his words. "You can have conflict without war. When a resource is scarce, either you fight over it or you share and comanage it. In the Jordan River Basin, there are other issues to fight about, so water may not be a cause of war in that way. But it will remain a source of tension. When you have an unstable situation and you add another source of tension, it can tip the balance."

Israel relies heavily on the aquifers beneath the West Bank, and its access to that water will remain important in negotiations. Israel is not likely to withdraw from the Golan Heights without a water agreement, and Syria will not give Israel a water agreement without its withdrawal from the Golan Heights. "If you want a world that works together, it has to be based on law," Elmusa insists.

Although Elmusa insists that water disputes *can* be settled among Jordan riparians, he doesn't promise it. "It's a common adage that negotiations are nonprincipled solutions." He smiles. "But if you don't negotiate to find a fair solution, the problems will not go away. On the Nile, Sudan wasn't happy with the 1929 agreement and so renegotiated another agreement in 1959, which contained a clause about future negotiations. It's a never-ending process. Obviously, if I'm going to negotiate, I'm going to be pushing for a fair deal, but finally you have to look at it within the whole gamut of issues, because water, however important, is not the only issue to be resolved. Finally it's a political judgment.

"In the end it is power politics," Elmusa insists. "The stronger riparians will compromise only to the extent they wish. But you will not have a peace agreement without a water agreement. This is where international law is important," he continues. "These are nation states occupying seats at the United Nations."

When Yitzhak Rabin signed the first historic peace accord, it was reported that he shouted, "Enough!" No matter how many wars are fought in the Middle East, there will still not be enough water to go around, so many believe that the Jordanian riparians ought to go directly to fixing the problems. There are a lot of things to be done here. They require technical fixes, money, good faith, and sharing. Much hangs on Israel's sticking to her peace accord agreements and to the principle of equitable utilization This will mean that every man, woman, and child in the land has the same right to water as every other. "Water was Allah's gift and no one had an ownership claim to it," says John Kolars. "Such may not be the case today." Water in the Middle East, more precious than oil and scarcer every day, will one day define the boundaries of life and war.

Along the Nile, the Tigris and Euphrates, and countless other rivers, man has tied his own beginnings to water. Many cultures have an innate reverence and respect for water as lifegiver. Although waterworks have been targets in wars, most peoples and even nations are hesitant to deny life's most basic necessity to others. The exceptions were those Serbs who lay waiting to shoot men, women, and children arriving at riverbanks or taps around Sarajevo carrying buckets and bottles, or Saddam Hussein, who diverted the lower waters of the Tigris and Euphrates to destroy the homes and livelihood of the Marsh Arabs. Few nations like to be thought so monstrous. But as tensions heat up over overdrawn resources, there is a burgeoning need for internationally accepted principles to help countries find rational ways to solve problems of water sovereignty and specific means of governance that take into account the tricky nature of water as well as its fundamental importance in sustaining life of every sort.

The nations that line the Nile, the Jordan, and the Tigris and Euphrates are members of the United Nations, the International Monetary Fund, and the World Bank, all of which can help through pressure and substantial incentives. When Syria and Iraq sent their armies to the border after Syria cut the Euphrates River flows, it took the intervention of both Russia and Saudi Arabia to quiet the two countries. But there is a limit to what outsiders can accomplish. Although they persuaded Syria to release more water and prevented a war, their achievement was not a lasting solution. Iraq, Syria, and Turkey will have to stumble along on their own path toward collaboration. The fact that they have continued to talk with one another rather than aiming more missile launchers at one another is, as geopolitical specialist Arun Elhance states, a remarkable testimony "to the power of hydropolitics to create interdependencies among riparian states."

"In the end," Elhance says, "even the strongest riparian states sharing international basins are compelled to seek some form of cooperation with their weakest neighbors."[172] He speculates that finally the most reluctant states will cooperate, because when water is scarce, the costs of noncooperation are felt on many levels. In the Indus Basin, David Lilienthal and the World Bank's Eugene Black led the charge to keep Pakistan and India away from their artillery. Because the two countries were anxious to find a way to settle the problems, and because of the intervention of well-meaning governments, a successful, enduring treaty was achieved.

"Hydropolitical cooperation may take a very long time to develop," Elhance concludes. "It may not necessarily lead to the optimal development and allocation of the shared water resources, may not satisfy or benefit all parties equally, and may not be possible without sustained third-party mediation and support; however, once achieved, such cooperation tends to endure."

Fadel Kawash, Palestinian water commissioner for the West Bank, said it more simply when he commented to a *New York Times* reporter: "A drop of water is the basis for living, and without it there can be no development and no peace."

Chapter 8

PRAYING FOR RAIN

*We live in the world's most technically sophisticated society, yet we
are now right back where we were three thousand years ago,
praying for rain.*

GARRETT WARD, *TEXAS DROUGHT*, 1997

I think I've learned more about water in India than in any other place on
the planet. The worldwide water crisis may be seen there in the starkest
possible terms: India's 20 percent of the world's population lives with
just 4 percent of its water—and even that percentage, often polluted
and already scant, is steadily declining. Indians venerate water. Hundreds of thousands of devout Hindus make their way to Varanasi each
year to wash away their sins in the river they call *Ganga Ma*—Mother
Ganges. They bathe in it, drink its sacred waters, reverently fill brass
pots and carry them home to bless their homes, cure the sick, comfort
the dying. But, as a recent morning boat ride I took along the bathing
ghats demonstrates, the same people routinely foul its waters: pristine
and blue at its birthplace in the high Himalayas, the Ganges at Varanasi is
brown and filthy, literally bubbling at some spots with untreated sewage
and effluents from nearby tanneries. The Ganges is an especially vivid
example of what may one day befall many of the world's waterways if
we are not vigilant and imaginative about applying new technologies.

India also shows how careful we must be about approaching present-day problems with the technological solutions that seemed most promising in the past. The success of Jawaharlal Nehru's visionary Bhakra Dam that provided power and irrigation to much of North India has not proved easy to replicate. "What happened after Bhakra was not good," the late B. B. Vohra told me. "Panditji [Nehru] was an advocate of big dams and big things generally. It became the fashion. States began to compete for large projects. They went too fast without adequate planning, without adequate finance, and now you can see the results."

Confirming Vohra's grim view of overbuilding are recent reports stating that many of India's large reservoirs are silting up at rates far higher than assumed when the projects were built, that the life span of major Indian dams is likely to be only two-thirds of design life, and that every dam built in India during the last fifteen years breaks rules meant to protect the environment.

Nonetheless, India has thrown herself into dam building with such vigor that one World Bank report called the country—now home to at least fifteen hundred big dams—"the world's greatest dam builder."[173] No country receives more Development Bank money for dam construction than India, though several of its gargantuan water-control projects are mired in controversy. The towering dam at Tehri, for example, which when completed will be the sixth or seventh highest in the world, rests uneasily on the edge of one of the world's most tremulous fault lines, the Central Himalayan Gap. Still more controversial is the massive Narmada Project—two enormous dams along with more than three thousand smaller ones that are meant to divert the waters of one of India's most sacred rivers from Central India's forested heart to arid regions of Gujarat State. In the process, thousands of square miles of precious forest will be obliterated, and hundreds of thousands of impoverished, landless tribals will be displaced.

When I spoke with the Indo-American engineer Ashvin Shah about the state of India and its water control, he told me that he believes the most critical issue in India's spurt of dam building may not be the dams so much as faulty processes of decision-making. He told me that in 1983, the late Prime Minister Indira Gandhi herself visited the Narmada Dam site and was appalled by the human toll it seemed likely to take. "Why is the government going ahead with this project?" she asked publicly. "My heart goes out to the poor people who will be displaced, but experts tell me there was no choice."

"In the face of self-appointed experts," observes Shah, "even a prime minister couldn't do anything else. The failure is at a democratic level. If the Supreme Court justices who allowed the Narmada to go ahead had opened the discussion up to anyone with a complaint, from the lowly farmer on, there would be accountability on the part of the engineers, the irrigation departments, members of Parliament, and ministers. A solution could have emerged that might have solved the problem.

"Technology is so important that it can either destroy mankind or save mankind," Shah continues. "Large-scale technical solutions should involve a consensus process. I'm talking about democracy. In my mind, whether or not to build a dam is so basic an issue that there is nothing to do except use a consensus process. Once you say a dam is the *only* solution, well, then, go into it, but first let's sit down and solve the problem. Technical processes are used to solve problems. Have you ever seen the medical profession put out a drug or a process without review? Why are we letting dams be built by politicians? The development of the Narmada River Basin ought to have begun with the development of the degraded land to prevent excessive runoff, and only then turned to dams." He points out that the state government of Gujarat, which will receive water from the Narmada Dam, has already built more than 131 large and medium reservoirs. "Since they have not solved the problems, what reason is there to think that more of the same will be effective?" he asks.[174] "It is high time that the World Bank and the Indian government recognize the technical infeasibility and economic unviability inherent in large dam projects in the degraded and heavily populated watersheds of India."

B. B. Vohra may have understood as well as anyone what can be done to forestall the crisis that has its fingertips at the country's throat. Tall and jaunty in one of the berets he favored, Vohra was a fearsomely outspoken man, but he was first of all a farmer who loved working his family lands. "It's a lovely sight, seeing the water gush out of your own well," he told me. Based on his experience as a farmer, Vohra—who had retired from government service in 1981—believed that India's efforts have been in the wrong places.

"Our preoccupation with big dam projects makes us neglect the real challenge in water management," he said, "which is the conservation of water as soil moisture and groundwater. There is a lot of money in big dams. I've had long arguments with the World Bank. A senior vice chairman once said to me 'From the bank's point of view, we would much rather deal with five projects of a hundred million dollars each than a hundred projects of five million dollars.' India also prefers big projects, but they are not the answer."

Vohra thought big, too, but in a very different way. He called for a countrywide soil- and water-conservation program. "We are still an overwhelmingly agricultural country," he told me not long before he died. "We must have water where rainfall isn't sufficient or timely. If it is not timely, you have to have some arrangement for irrigation. Where water is scarce, we must have arrangements for conserving it. We are a poor country. We haven't the ghost of a chance if we don't make the best use of our natural resources."

India also provides examples of ordinary people who are trying to do just that, working together to solve water problems from which the rest of us would have fled long ago. Just a few years back, the villages of Dholera District along the Gulf of Cambay in Gujarat, India, were literally dying of thirst. The thick green forests that had once surrounded them had long since fallen to the woodcutter's ax. There were no permanent rivers, no lakes, no available groundwater, just endless cracked earth and overlying layers of salt.

Dholera's people could manage during the monsoon, but afterward, as their little ponds dried up, sentries were posted to guard the little that remained, and village women would wait all day for the sporadic visits of water tankers, which, when they finally arrived, doled out just two quarts per person. Men fought over small pits of briny water, rode miles in bullock carts in the hope of harvesting a few precious drops from leaky water pipes. Village life was fracturing. The people of the district were forced to go on the road with their cattle for eight months a year, leaving behind only the very young and the very old.

In 1985, seeking a way out of such trouble, the people of the village of Rahatalav decided not to migrate. Instead, they resolved to work together to dig a pond the size of a football field, then lined it with heavy plastic to keep monsoon water from seeping into the ground. A plastics manufacturer from Baroda instructed the villagers in how to fashion well-graded slopes, build sand filters, and add pumps for drinking water. The women of Rahatalav were especially heroic, organizing, excavating, grading. The well-engineered pond was ready for the 1987 monsoon, but somehow the rains passed it by. Acting quickly, the villagers dug a two-mile channel from a place where rain *had* fallen to their pond, and it filled beautifully. In the years that followed, when villages all around them dried up, the people of Rahatalav welcomed their neighbors to drink. Since then, at least seven more villages have built ponds of their own. The end of the story is simple. The people of Dholera District have water and they take care of it.

Dholera's life-giving ponds are not new: three thousand years ago nearly every Indian community had its own water tanks. In parts of South India, ancient tanks sometimes line up back to back like railroad cars, so that overflow spills from one to another. But, sadly, the tanks, some bigger than city blocks, have fallen into disrepair and the solutions of the past lie broken and overgrown with grass grazed by scrawny cows. In the Ramnad District in Tamil Nadu, where water lords now control every well and water hole, eighteen hundred tanks are still in use out of the original six thousand.

Convinced that rain is the base source of our water and that it should be caught and stored as close as possible to where it falls, a sixty-four-year-old Indian newspaperman named Shyamji Antala has led a one-man crusade to restore water sources and recharge wells. Antala, having identified rainwater collection areas, has taught Gujarati villagers to send rainwater underground into deep wells and aquifers for storage using simple cement pipes and tanks to filter sediment. Around Gundasara Village, 210 out of 277 dry wells are full again, and the crop returns are excellent. Antala, who has helped restore over 350,000 wells in five years, now sees his work beginning to be duplicated in other Indian states.

Engineer Ashvin Shah, who has scrutinized Antala's work with a wary technical eye, concludes that it can provide genuine answers for India's water needs, answers far better suited than a gigantic dam project like the Narmada, which was never subjected to rigorous technical assessment. "Rainwater-harvesting and groundwater-recharging projects are technically and economically viable even in semi-arid Gujarat," says Shah. "Ordinary villagers have responded in thousands to make it a peoples' movement." The revitalization of village water resources has been called a silent revolution in eastern India, as thousands of villages in dry regions have restored their well-being and prosperity. Shah feels that rainwater harvesting in conjunction with surpluses from India's surface reservoirs mean that all of India can have water and that this understanding has meaning for the entire world. "There is enough water if we plan for it," he steadfastly insists.

WE WANT COMFORTABLE, warm places to live, clean water flowing from our taps, plenty of electricity, tomatoes and strawberries in January, transportation to carry us wherever we want to go—all the good things that technology promises. And, oh yes, we'd also like clean air, pristine lakes, and unsullied wilderness. With so many people sharing the planet, it's unlikely we can ever have all these things at once. But to the extent we can have it all, engineers will help make it possible.

They have already had an astonishing impact on the planet: by consolidating so much water in deep reservoirs they have actually sped up the earth's rotation, according to one estimate, shortening each of our days by one eight-millionth of a second and, by creating a new wobble, are estimated to have moved the North Pole twenty-four inches closer to Canada since 1950.

Till now, the preoccupation of most technical experts has been to get things built to solve a specific problem: getting irrigation water to farmers' fields in the Punjab; protecting lives from the sea in the Netherlands; electrifying the Tennessee Valley. Many of the structures worked, but they were all too often accompanied by unintended consequences, by mistakes and shortsightedness. Benefits were overestimated while economic, environmental, and social costs were underestimated. Because we had the means of ever more powerful technology, we used it extravagantly, always expecting that we could locate another water source or fix whatever problems we created. We began the mighty engineering works of the twentieth century in environmental ignorance and ended the century in environmental crisis.

For this, engineers often get the blame. There's nothing new in that: according to William Willcocks, when one of the Tigris levees broke in the seventh century A.D. the local king threw four hundred engineers and their supervisors into the breach. And the critics often get things wrong too: I'll never forget a boat ride across a blue lake alive with waterfowl in the company of an especially impassioned Indian conservationist. "I *hate* dams," he bellowed conversationally, unaware that he was floating by the grace of a stout but unobtrusive little dam at the lake's eastern corner.

In fact, for the most part engineers have been only as good—or as bad—as the tasks they were given by the rest of us; the Aral Sea was drained because Ukrainian cotton farmers wanted more land for their crops; farmers and housing developers around Florida's Everglades pressured the Army Corps of Engineers into ruining much of those precious wetlands; the Yangtze is being dammed because China's leaders don't know how else to stop its flooding or provide the power their country's soaring population demands.

Some engineers have always taken a longer view. Mark D. Hollis, for example, Chief Engineer of the U.S. Public Health Service, publicly warned of the danger of dumping chemicals into American water almost a decade before Rachel Carson published *Silent Spring*. "If we had asked our engineers to look into the question of overall protection of our resources," the engineer Samuel C. Florman has written, "they would have been happy to do so; many of them were anxious to do so and said as much."

I asked Pedro da Cunha Serra, president of the Portuguese Institute of Water, if the engineers he encountered in the field were changing. He was certain that they were. "Nobody wants to spoil the environment," he said, "even if we have different points of view. I believe that civil engineers are quite aware of the need to care for the environment." I pursued the same question with many of the engineers I talked with while researching this book. Most did seem to have environmental concerns. "The engineering community is aware of the fact that projects in the past were realized without sufficient thought about long-term effects," Dutch engineer Gijsbert Te Slaa said. Engineering education, he pointed out, like that of today's doctors, demands mastery of an ever more daunting aggregate of technical complexities. But big projects also now require input from apparently unrelated fields—sociology, health, diplomacy, political science. "An integrated development philosophy has become important," Te Slaa continued. "You must always look at what the project will do to the environment and to the mostly poor people who are most directly affected by it."

Who else but engineers can re-engineer engineering disasters? Who else will come up with fresh solutions to our most pressing water problems? While low-tech solutions are especially helpful in developing countries, Slaheddine El Amani, director of a Tunisian land-and-water-resource center, has asked sensibly enough why we cannot combine "the best of the old with the best of the new."

Some of the new is simply astonishing.

To begin with, we can make it rain. When we inject silver iodide into clouds, noncrystallized, dispersed liquid droplets are given a nucleus around which to crystallize and form ice. As the ice becomes heavier, it begins to fall, picking up more droplets along the way until, heavier and heavier, it tumbles into warmer air, melts, then falls as rain.[175] Given the right weather conditions—which means lots of clouds—this technique can boost rainfall by 30 to 40 percent. Cloud seeding has been used in parts of Texas and Utah for almost thirty years, and operators can now trace chemicals with high-speed analyzers and computers to predict where and to what degree seeding will work. Even more ambitious, Electrificación Artificial de la Atmosfera and the National Autonomous University of Mexico are experimenting with a weather-changing antenna system that ionizes the atmosphere itself, generating nuclei around which water vapor can condense in order to alter air currents and make rain without anyone having to leave the ground.

Miraculous as it seems to the layman, cloud seeding is merely an "episodic enhancement," tapping available moisture.[176] The alchemy of desalination, however, offers us the oceans themselves. Aristotle explained how it works in 300 B.C.: "Salt water, when it turns into steam, becomes sweet and the steam does not form salt water when it condenses." Distillation is the most common and time-honored method: when salt water is boiled, the steam can be captured, leaving salt behind. In the membrane or "reverse osmosis" processes, seawater is forced through semipermeable, ion-specific membranes, which sieve out salt ions.

Desalination may be the only option in dry and desperate places like Gaza. But it uses a lot of energy, which makes it useful for producing drinking water but prohibitively expensive for making water to pump onto fields. There are thousands of desalination plants around the world, most of them in the Gulf states, where energy is cheap. (In fact, much of the water used for petrochemical production comes from desalination plants, which in turn are run by petroleum.) The United States is the world's next largest builder of desalination plants after Saudi Arabia and its neighbors. Recently, waterfront towns in New Jersey built plants to desalinate aquifer water made brackish by seawater intrusion. Soon, America's largest membrane plant, working alongside a power plant in Tampa, Florida, will pump 25 million gallons of water a day at a consumer price of $2.08 per thousand gallons, comparable to current water costs. Israel, facing shortfalls of 300 million gallons daily, is contemplating several large plants on the Mediterranean, intended to yield large quantities of water at half the cost Israelis now pay—and simultaneously help solve some of the country's political problems with her neighbors.

Harvesting the rain is done in both new and old ways. In its older incarnations, simple microcatchments are used to hoard water around plants or trees or even stone walls, like those in one Burkina Faso village that were used to hold soil and rainwater and that increased crop yields by a third. In a new twist, soil can be treated with water-repellent or surface-binding agents to push water where it needs to be. In Abu Dhabi, scientists working with a company called Light Works have built a large oceanfront greenhouse that sucks moisture out of desert winds through an ingenious combination of distillation, air-conditioning, and photosynthesis. The process provides plenty of water to grow vegetables and as much as ten gallons of water a day to spare for each square yard of greenhouse. Storing water underground, as Saudi Arabia has done with the use of recharge dams, is a longstanding method of storage that minimizes loss to evaporation and refreshes aquifers. By using deep well injection to push Colorado River water underground, the Las Vegas Valley Water District has raised aquifer levels by as much as eighty feet.

Conservation is an old idea, a substantive source of water, and science has found a host of new ways to do it. Simply fixing leaky pipes can save a lot of water. While England's water companies lose 30 percent of their water to leakage, in poorer countries losses run from 40 to 60 percent. England's Defense Evaluation and Research Agency in Dorset is looking at finding water leaks with the sonar technology used to locate submarines.[177] In Denver, city fathers worked out that it was less expensive

to subsidize water-saving toilets and showerheads than to build the Two
Forks Dam. In a conservation project in San Jose, California, begun in
1990, fifteen companies reduced annual water consumption by more
than a billion gallons a year, enough to meet the needs of 7,000 house-
holds, and at the same time saved around $2 million a year.

Catching used water and using it again is another source of water
made available by technology.[178] Industrial recycling systems are especially important, particularly in operations that use a lot of water—
paper, chemicals, petroleum, primary metals like steel, and electronics,
where very clean water is needed. "Over the coming years, industries
that consume large amounts of water will be economically vulnerable," I
was told by Paul Pimentel, an industrial-water-systems designer who
feels that severe water conditions indicate a tremendous demand for
water-supply technologies. "With the marginal wholesale price of water
in the United States expected to triple in the near future, investments in
conservation and reuse by industry are expected to pay off very quickly."
Companies like Xerox, Hewlett-Packard, and Intel have all introduced
in-house water recycling. Xerox, which spent $50,000 in conservation
equipment in one location, saved $38,000 the first year.

"Most companies putting in water conservation methods will achieve
payback in two to three years' time," says Pimentel. "But where systems
are designed into new plants and in high-value industries—such as
semicon-ductor manufacture, where water reliability is important—the
savings are really substantial." It takes 3,000 gallons of ultrapure or
deionized water to make a silicon wafer. In the 1990s, Pimentel assisted
in the design of a biotechnical facility with cost savings of a million dollars. In thermal and nuclear power plants, which use large amounts of
water for cooling, Pimentel says it is possible to recycle 98 to 100 percent of the water.

An in-house recycling-and-waste-treatment system developed by
Danish catalyst manufacturer Haldor Topsøe produces ultraclean water,
saves both water and energy, and doesn't contaminate water sources. I
talked to the company's founder, Haldor F. A. Topsøe, in his sparkling

plant near Copenhagen. "When operating a chemical plant, you always have effluent and you always have to clean it up," he explained. "We've developed a combination of membrane and high-efficiency evaporation technology for recycling most effluent streams. We use it in plants where we handle substantial quantities of heavy metals. It allows us to recycle one hundred percent of our wastewater." The clear blue, dancing stream behind the Topsøe offices seemed to confirm the chairman's view.

Recycled sewage water can be a critical water source for municipalities as well. Hundreds of American cities reuse treated wastewater for golf courses or traffic strips.[179] The cities of Tampa and San Diego have refined wastewater until it's clean enough to drink.[180] Ashvin Shah estimates the costs of the process to be half that of desalination. "Recycling municipal and industrial wastewater combined with increased efficiency of water use in agriculture can solve water conflicts around the world," he muses happily, "leaving policy makers at political levels free to quarrel about other issues."

POLLUTION HAS, of course, denied much of our water to us. World water experts meeting in Japan in 2001 concluded that more than half the water in our lakes and reservoirs—that's 90 percent of the world's liquid fresh water—have become polluted, putting a billion people at risk. But even here, science and technology offer new remedies. A Cornell University team, for example, is working with genetically modified bacteria that will clean up mercury; German researchers have discovered bacteria that mop up toxic chemicals like solvents, paint, and glue; a New Mexico firm, Second Chance Water, claims that it has converted oilfield brine into drinking water; and the University of Southern California School of Engineering is already creating swarms of tiny robots to monitor toxic algae and other dangerous organisms in the oceans. Anaerobic waste treatment, enzymes, bacterial bioreactors, and remedial treatment of groundwater are just a few new methods of cleaning up some of the messes that we've made.

"We are approaching some very difficult situations in the balance between water and people," writes Russian water expert Genady Golubev. "The answer is better use of existing water." Although there are many ways of using our water better, some of the most important solutions to water problems will be found in the hands not of hydrologists and engineers but of economists.

THE ANSWER TO water shortage is not always more water. Sometimes economic adjustments are important. But applying monetary value to water is an emotional, even confrontational, idea. I talked with Florida environmentalist Joe Podgor about the perils of trading water for money. "Water is like air, without it you die," says Podgor. "It is the only thing that the government gives you without which you cannot live. And to say to me that it is priced too cheaply is like saying that air shouldn't be free anymore." Podgor believes that the government has an obligation to provide us with clean, plentiful water at the lowest possible cost. Anything less than that, he says, is "a violation of my human and civil rights. So when they say it costs too much or costs too little, or we should levy a surcharge, they're trying to say that you should breathe more shallowly."

In many places on the planet, governments fail to meet that responsibility, and a quarter of the world's people do not have access to safe drinking water. Under these circumstances, money often dominates its distribution. In Cairo, the poor pay vendors 40 times the real cost of delivery; in Karachi the figure is 83 times; and in parts of Haiti, 100 times, or a third of residents' income. Poor farmers without water in South Asia often surrender a large part of their crops to those who own pumps.

There is no doubt that one of a government's prime obligations should be to get clean water to its people, but someone has to pay for its delivery. Average municipal water recovery costs around the world are 35 percent of the expense of supplying it. Governments commonly subsidize the water, making up the shortfall by not maintaining storage and delivery systems or bypassing the poor. In Jakarta, the 14 percent of mostly middle- and upper-class residents who receive piped water from the city system pay 9 cents a cubic meter, not enough to cover anything close to the real costs of supplying water. This means that there are no citywide delivery systems, and poorer people must buy dirty water from street vendors at 60 times its cost. But in another Indonesian city—Bogor—when water fees were tripled, consumption dropped 29 percent and water was made available to hundreds of thousands who had not previously enjoyed it.[181]

Where water is underpriced, it is often overused. It is generally true that people who pay for water use it more sparingly. Water use has dropped persistently in the United States over the past twenty years, at least in part because of higher charges. Yet, industrial water-consumption in Germany, where water prices are still higher, is only about half that of the United States. In Sydney, raising usage fees meant that consumers saved enough water to avoid building a new dam. Some economists have proposed pricing water according to its expected future supply cost, or marginal cost pricing.

Valuing water as a commodity takes creative thinking. In Irvine, California, when water users conserved water during droughts, fees for usage dropped so low that the water district couldn't afford to do business. So the California district developed a rate structure that guarantees domestic users enough for basic needs but penalizes those that dump water as if there were no tomorrow.[182]

Professor Tony Allan, a specialist in Middle East and North African resources at the School of Oriental and African Studies in London, believes that the most critical answers to water shortage are to be found in political and economic management. Allan, a large, authoritative man with a shock of white hair, argues that believing, as so many people do, that water should somehow always be free is potentially lethal because it escalates the expectations of farmers and prevents officials from making rational judgments about how water is priced and allocated.

"Much of the trouble in arid places like Arizona and California lies with stupid water budgets that support a lot of irrigation," Allan says. "If you decide in hot country to allocate water to agriculture, irrigation will swallow up most of the available water." Agriculture represents only 10 percent of California's economy, he notes, but commands 80 percent of its water. It's a point that Allan has made over and over again. "Food should be grown where water is free," he says adamantly. "Absolutely free. In a hot country, it is never free." By the time water reaches the end of Israel's National Water Carrier, he explains, it costs about a dollar per cubic meter. Even a nonthirsty crop growing on a hectare requires ten thousand cubic meters of water. "Ten thousand dollars for a hectare? No one should export citrus and other crops from a water-short country, because if you figure in the real cost of water into it, you will lose money *every time*." Economists agree. In northern China, planners calculated industrial water to be worth sixty times as much as agricultural water. In California, industrial water has been valued at sixty-five times the same water used for crops.

Because of the overwhelming cost benefits, Tony Allan predicts, in the future most countries in the Middle East will take water out of agriculture and put it into industry and domestic use. Saudi Arabia has already curtailed grain exports in order to save water. And yet, I asked Allan, what will happen as that agricultural capacity disappears? Who will feed those people? "Well, that's the question to ask." He smiles broadly. "Water is a natural resource, which enables you to generate wealth, and if you can generate wealth in industry you can then buy food. It's a much more efficient way of using water. Egypt is short of nearly half of its food because it doesn't have enough water. But it is not starving. If you were to have said twenty years ago, 'Could Egypt possibly manage growing only half its food?' people would have said it must lead to war. But it hasn't. Egypt can deliver the entitlement to food, even if it has to import it. And that's the remarkable story—not the fact that the pressures are mounting, but that political and economic adjustments are being made."

Such adjustments are not for the faint of heart. They necessitate steadfast decisions by our politicians regarding just where the water is going to go. This may mean taking it away from people who have been using it their own way for a long time. "In the business of managing water, the political decision to allocate is the important one," says Tony Allan. "Most people don't want it to be a political one, because it raises all sort of problems with farmers, developers, industry, and on down the line. So it's best to pretend the past allocation is the only allocation. You can understand why politicians want to look the other way." The battle for reallocation is being fought out in many dry places, few more visibly than in the American Southwest, where politicians have looked the other way for a long time.

In December 1997, U.S. Secretary of the Interior Bruce Babbitt announced at a meeting of the Colorado River Water Users Association in Las Vegas that the federal government would develop a regional water market, beginning with interstate sales of water from the Colorado River. This makes it possible for those already holding the rights to a certain portion of water to sell it rather than use it. Arizona can sell to Nevada. Farmers can sell water to cities or to electric utilities instead of pouring it on crops.

Although creating a regional market in the American West has allowed some water to get where it is badly needed, the basic insanity underlying Western water allocation remains unchallenged, particularly along the Colorado. One-sixth of Colorado River flow has been allotted and delivered inexpensively to the fortunate corporate farmers of the Imperial Valley Irrigation District since the early part of the twentieth century. *The Economist* summed it up: "The advantage of water transfers is that they begin to bring the price mechanism to bear without challenging the status quo. This is not a real market because it consists of one set of owners who get water free selling it on to another set with no such luck. But in the mad, mad world of Californian water, it constitutes a great leap forward."

Water markets function in a variety of ways, the purchase of land with water rights being the simplest. Leasing or selling water rights irrespective of land is more complicated, and the selling of water in the American West is sometimes a clunky business, slowed by a briar patch of legalities in which rights are almost always dependent on a variety of federal programs.[183] In the early nineties, for example, a Texas development company purchased forty thousand acres of farmland in California's Imperial Valley intending to sell the water used to irrigate that land to San Diego County. But before a drop was sold, they discovered that the water rights were actually owned by the Imperial Irrigation District. The district then decided to sell that same water to San Diego County only to find that even with their rights in hand, a lengthy court battle was required to make the water transfer possible.[184]

To facilitate water marketing and as a hedge against drought, some states have worked together to create water "banks." In 1999, Secretary of the Interior Bruce Babbitt made it possible for the state of Nevada to store surplus Colorado River water in Arizona aquifers. Today, the two states operate a joint accounting system of deposits and withdrawals, much like any conventional bank.

Sometimes measures meant to straighten things out go badly awry. Early in the 1990s, California and Idaho created state-run drought banks to buy excess water and resell it to waterless farmers. In Butte County, California, third-generation farmer Carl Starkey watched his wells, like those of his neighbors, drop 40, 50, then 85 feet, as landowners in adjacent irrigation districts pumped water from shared aquifers to sell to California's Drought Water Bank. "Some of the sellers would drill a huge hole," Starkey complained, "put in a gear head, back up a D-9 tractor to it, and pump five thousand gallons a minute." The sellers cleared as much as $5 million in three months while Starkey and other farmers lost crops and large amounts of money drilling deeper and deeper wells in futile search of their own water. More than one farmer lost his farm.[185]

Despite the apparent inequities, the rising value of water means that money-driven markets are asserting themselves all over the world. Such markets may afford little or no protection for public health, wetlands, wildlife preserves, or communities without economic power. "There may be other public interests that have to be addressed concerning water," comments Democratic Californian Congressman George Miller, apart from "simply allowing money to move it." California history provides a clear example of what he means: in 1913, the city of Los Angeles completed a system of aqueducts and siphons and tunnels that brought water to the city from the Owens River Valley, 250 miles away. The result was mushroom growth for Los Angeles and disaster to the farmers of the valley. "In its rawest form, Owens Valley was water marketing," Miller says. "It may be that the city of Los Angeles had a higher and better use for that Owens Valley water. But I'm not sure we want decisions today made that way throughout the West. If you end up with pure water marketing, all the water in the state of California would end up between Santa Barbara and San Diego.

"It's one thing to have water contracts governed by statutes that provide certain rights in drought conditions to recall a portion of the water," Miller continues. "It's another thing to let those situations be governed by contracts that may not take droughts into account."

The country of Chile has been especially successful with water markets, in 1981 having put in place a system of water rights independent of land that can be sold, traded, or allocated at prices regulated by the market. In order for this kind of market to operate, ownership of water rights must be absolutely clear. Chile's advantage was that beginning in 1966 the state appropriated all water rights, so it was able to start from scratch. Henceforth, under Chilean law, allocations were not attached to past or prior use.[186]

Furthermore, water must be extracted so as to not damage the resource or harm third-party rights. An entire chapter in the code concerns underground water, establishing an area of protection for grants of groundwater where wells or pumps cannot be used. In case of dangerously depleted aquifers, the General Directorate of Water (DGA) can restrict access and will move against those committing water violations. When a drought zone is declared, if users don't agree on distribution, the DGA can distribute water for public use. The agency must also authorize any major water-related infrastructure, such as big dams or large canals.

Ninety-nine percent of the Chilean population in cities and 94 percent in the country now receive water, compared to 64 and 27 percent, respectively, in 1970. Cities purchase water from farmers who have adopted water-saving irrigation technology so that they still have sufficient water for farming and enough left over to sell. Agriculture is doing well. Land use has shifted from cattle and grain to tree crops, such as fruit and grapes, and the country is a major supplier of fruit and wine to the United States. Water markets can even level inconsistencies in water value. Since water in Chile has been channeled to the best economic uses and distributed efficiently, it has also been priced more accurately. Chile's successes brace arguments for more valid pricing of water services, arguments that we'll be hearing more about in the future.

Tony Allan reminded me that adjustments are being made. While the past decade has seen water become scarcer and more valuable, there has been an increasing recognition of that shift in new regulations. State officials in Kansas recently agreed to penalize people who overpump wells. California state authorities are discussing curtailing water deliveries to Imperial Valley farmlands to protect the Salton Sea from becoming too salty to sustain wildlife. The New Mexico legislature has enacted a landmark law, devised by State Engineer Tom Turney, that forbids developers to build unless they have water rights to accompany their plans.

Even the most exalted political idealism is no substitute for an intelligent scrutiny of engineering projects. "The obvious problem with letting the few make decisions for the many is that the many, if they knew more, might want to do something different," writes the engineer Samuel Florman. "Since decisions about water are made in the halls of government, then all of us should know as much as possible about the source of our water and how control works. A knowledgeable public will not expect to resolve each technical issue by analyzing evidence, but will see to establish a fruitful relationship with its experts—and its politicians—a combination of trust and suspicion, respect and obstinacy, calculated to best translate social objectives into technical decisions."

Methods of addressing water problems, both new and old, are being facilitated by public involvement in water decisions and its subsequent effects on water conservation. People across the spectrum of environmentalism, politics, and development are finding ways to work together. Coalitions of wide varieties of organizations have never been more innovative or more persuasive, and there have never been more of them. In Washington State, the Environmental Defense Fund, Washington State Department of Ecology, the Yakima Indian Nation, and the U.S. Bureau of Reclamation worked together to purchase irrigation water rights from farmers and leave the unused water in the Yakima River to restore the river and its wildlife and add hydropower downstream. Along the Feather River in Northern California, fishermen, farmers, landowners, the Army Corps of Engineers, and a variety of other folks cleaned up 38 creeks and brought erosion under control, bringing millions into the local economy at the same time the fish came back. In Divide, Montana, the members of the Big Hole Watershed Committee represent a variety of factions from ranchers and outfitters to government agencies. They all care about a healthy river and are willing to make investments in natural resources that will in turn stabilize local economies.

These fixes, however, remain local and piecemeal, while problems often traverse state lines and affect whole regions. In the Murray-Darling Basin—a vast territory, as large as France and Spain combined, that sprawls across parts of four Australian states—a relatively new governmental organization, the Murray-Darling Basin Commission, is trying to find comprehensive solutions to the problems faced by all its citizens, many of the same problems that plague people living in river basins all around the world.

THERE ARE TWENTY major river systems in the Murray-Darling Basin, most of which spill out of the Great Dividing Range, a labyrinth of rivers and wetlands and billabongs that act like the circulatory system of the body. Bottle trees swell strangely out of golden grasslands grazed by sheep. Plots of farmland are laid hard next to one another, orderly chunks of green and brown trailing into the distance. This region is the jewel in Australia's agricultural crown. Its farms and agro-industries yield half of Australia's farm production. The basin is home to half of Australia's sheep, more than half its orchards, and a quarter of its cattle. It also holds three-quarters of the nation's irrigated land, where every dollar in profits begets five more dollars in the community at large and every on-farm job creates four more.

All this apparent prosperity came at a fearful cost. The first settlers assumed that Old World–style irrigated farming could somehow be made to work in the harsh Australian landscape. Old, fragile soils soon crumbled under the piercing hooves of alien livestock, were blown away in the wind, or were washed away by rain. Wholesale tree cutting and land clearing, coupled with the heedless digging of canals and the building of farm dams—there are said to be a million and a half of them in the basin—combined to begin a process of devastation. Farmers eventually found themselves paying the price in rising water levels and salty white tracts in their fields. Wetlands, the cleansing organs of the river system and its most biologically exuberant areas, were ravaged— drained, built over, polluted, choked with trash. Water diversions—in addition to all the farm dams there are eighty-four substantial reservoirs in the Murray-Darling system—have put an end to normal flooding. Only a third of the water that enters the Murray-Darling River system now reaches the ocean.

As early as 1863, farsighted Australians had begun to talk of somehow finding a way to manage the whole basin, but it was not until 1915 that the governments of Queensland, New South Wales, Victoria, and South Australia actually signed an agreement to share waters equably. And not until 1987, after South Australia sued New South Wales to stop it from issuing more water allocation licenses, was the Murray-Darling Basin Commission finally given the power to make decisions concerning the "planning, development and management of the water, land and other environmental resources" of the entire region.

The decision did not come a moment too soon. State and local agencies had clearly long since been overwhelmed. Loss of production due to land and water degradation in the basin is still conservatively estimated at a stunning $390 million yearly while pollution from farms, factories, and city sewage on the upper Murray and Darling rivers has been carried deep into South Australia. In 1992, as a result of this fouling, the largest river bloom of blue-green algae ever seen anywhere clogged the Darling River, killing fish and river organisms, poisoning cattle and sheep along its banks for more than 600 miles.

"There is no more powerful stimulus than the distribution of wealth to focus the political, bureaucratic, and community minds on issues," says Don Blackmore, the commission's youthful CEO. The commission's first job was to identify what was wrong and why, through mapping, modeling, reviews, and exhaustive studies. Then it moved on a whole range of fronts, seeking always to involve everyone affected by its decisions—from innocent-eyed schoolchildren who were taught to plant trees and sing songs about cleaned-up rivers, to the hard-nosed farmers and businessmen.

Realism is an essential element of the commission's work. It recognizes, for example, that it is unlikely that either salinity or waterlogging can ever be entirely eliminated, but a series of imaginative programs have slowed both appreciably. Limiting how much land can be used to grow rice has helped; so have deep drainage, insistence on sprinkler or drip irrigation, tree-planting, and putting in place an innovative system of salt credits, under which farmers willing to ante up the cash can buy the credits of others not using their allotments. The money received is, in turn, plowed back into schemes that further reduce the impact of salts. In one such scheme, salt water is pumped out of bore wells and piped fifty miles away to a natural evaporation basin; thanks to another, which retains irrigation drainage on farms, district groups are able to extract four truckloads of salt out of the Murray every day. Over all, such interception schemes remove upward of twelve hundred tons of salt out of the Murray every day.

The commission has also taken on pollution. To deter algae blooms, for example, scientists have identified the sources of toxins, enhanced sewage treatment, constructed artificial wetlands to absorb some pollution, cut back on fertilizers—and done everything they could to get choked rivers flowing again. Dairies were encouraged to store wastewater through the winter so that it can be reused as fertilizer in spring. Australian Newsprint Mills, a major paper manufacturer, which once poured tons of polluted water into the Murray, now treats its wastewater instead, using it to water its own tree plantations. Enlisting the help of hunters and fishermen, they've restocked native fish and coped with an infestation of European carp, which crowds out local fish.[187] They've re-vegetated riverbanks to prevent erosion, prohibited livestock on the banks, and planted a billion trees.

At the heart of the metamorphosis of this big-money region is the basin's governmental format. The Murray-Darling Basin Commission is headed by a Council of Ministers from the basin's four states and the Commonwealth. Each of the river systems within the basin has its own catchment committee supported by community groups. There isn't an answer for every problem in these troubled plains, but the Murray-Darling Basin Commission's framework allows every question to be tackled by a cross-section of concerned parties.[188]

One of the most onerous jobs the commission has taken on is the imposition of a cap on extractions and a moratorium on future diversions, which simply means that no one can take water out of the system without the commission's approval. While farmers and developers won't be happy about being denied as much water as they wish, it will prevent overuse and restore the river system's ecological health. The commission has set aside water allotments for wetland replenishment and introduced artificial flooding along the rivers to allow wildlife to finish breeding seasons and to allow the propagation of finicky red gum trees. The groundwork for allocations was established in a 1995 water-use audit, which enables the Murray-Darling Basin Commission to understand where water goes and how it is used, and to assess the right balance in the diverging needs of humans, industry, agriculture, and rivers.

What's been accomplished here? The commission is constantly assessing and redefining its task, and Blackmore doesn't overplay its successes or underestimate what remains to be done. "Growth in water use has been halted," he has written. "And we are now struggling to determine what a sustainable river is and what flows are needed; we have a strategic approach to salinity management; we have the early makings of an integrated-catchment management approach; and we have just scratched the surface on how to change farming systems and reintroduce vegetation back into the landscape."

Blackmore's modesty aside, it seems clear that when it comes to management of river basins, the Australians are well ahead of the rest of us. In the United States, for example, the Environmental Protection Agency recently issued a report affirming that the watershed approach offers the best hope for protecting and restoring the nation's river basins. But it then went on to say that private citizens must lead the drive to reverse harmful effects. Citizens' groups can have an enormous impact, especially when they pressure the government into doing its job. But, without mounting forbiddingly costly lawsuits, they don't have the authority to move county, state, or federal agencies to action. They can't work across state lines, penalize polluters or overpumpers, resist vested interests, or fund the exhaustive scientific research needed to understand a river basin. They can't allocate water or cap its use. To do those things across an entire watershed, government—at both state and national levels—needs to lead.

When Don Blackmore speaks of the "overlay on the federal system" that the Murray Darling Basin Commission has created in Australia to manage a shared, limited resource, he is deeply concerned that it create "a sense of mutual obligation strong enough to overcome a narrow self-interest."

"There are no 'one-size-fits-all' solutions to water problems," Black more says of his organization's ongoing work in Murray-Darling Basin. "The journey has just begun. But we will succeed." The integrated management he helps to run remains the most comprehensive in the world, continuing to tackle water problems on an unprecedented, all-inclusive scale. The lessons the Murray-Darling Basin Commission learns along the way should be of immeasurable help in solving the all-too-familiar dilemmas encountered in river basins everywhere.

Epilogue

THE EVERGLADES

It's a race between man's intelligence and his stupidity.
I don't know which will win.

MARJORY STONEMAN DOUGLAS,
INTERVIEWED ON HER 100TH BIRTHDAY, 1990

Right now a gallon of drinking water in my city costs more than a gallon of gas. I buy that gallon of water anyway, because of cryptosporidium and E. coli, visitors I'd never heard of until recently. I now boil my tap water in New York City as assiduously as do my friends in urban India. Even in America, the rules have changed. Many of us are revising our ideas about water. I've been reading about water in newspapers for more than a decade and have seen a gradual but definite evolution in water thinking.

"Our whole society was built on the notion that we could and must control nature, that we must master our circumstances, technologically," environmentalist Roland Clement said to me not long ago. "But natural systems are the consequence of a long evolution, and ecology is teaching us that we must first understand these systems to see how far we may modify them for our benefit without disastrous consequences. This is a new point of view that arose with ecological science, that world systems have a functional reality of their own and that if we push them too far, the systems will either break down or backfire."

Can we manage a dwindling resource? Yes. Can we stop the warming of the planet? Yes, we can, but we've got to pay attention. We need to think about where our water comes from. We have to let our politicians know that our water matters to us. Standing in dry fields and boggy swamps around the world, talking to farmers, engineers, environmentalists, development workers, and ordinary townspeople, I have become painfully aware of the urgency of ensuring water, as vital as blood, to quench human thirst and hunger without damaging the water-based systems on which we depend. We need to conserve, protect, assure best use and reuse, recycle, apply technology, and make politicians pay attention to the larger issues.

"This is not a 'save a bird, hug a tree' issue," says my friend Joe Podgor who, along with other South Florida residents, has fought for clean, ample water for Miami and for the Florida Everglades. "Do you like coffee in the morning? Would you like to drink your cup of coffee? Do you want to be able to take a shower or would you like industrial waste coming from your faucet? Land-use control in the recharge area is the way to keep the water clean in the well field. If you don't protect that well field, what they spill on Interstate 75 on Monday will be in your coffee on Tuesday, coming right through that spigot.

"You find out where the water that you drink comes from and you keep it clean," says Joe Podgor. "It's that simple. Its very difficult to get people interested in water as a global or even regional issue, but if you bring it down to the kitchen sink, we all do get the general picture."

THE STORY of the long struggle that Podgor and some of his friends have helped wage to restore Florida's water demonstrates both the terrible damage that can easily be done when men act to alter their world without fully understanding how it works, and the no-less-remarkable power that determined citizens have to remedy those mistakes before it is finally too late.

Florida is surrounded by and saturated with water. Water pulses through its labyrinth of waterways and spills in excess from both its coasts. But at the state's southern tip, the Everglades—one of the world's richest subtropical wetland wildernesses—is dying of thirst. It is a man-made drought, the result of hubris, faulty engineering, and an almost-willful misunderstanding of nature. The fight to reverse what human beings have done there has become the biggest political battle and engineering initiative in American history.

In the late nineteenth century, South Florida was still mostly possibility, but politicians, speculators, and farmers had already begun to see in the peaty muck of the low-lying wetlands around Lake Okeechobee agricultural empires and easy money. The first canals were cut across the state to drain boggy land for farms in the 1880s, but the actual draining of the Everglades did not begin until 1905. It was the favorite cause of a charismatic new governor, Napoleon Bonaparte Broward. A former gun-runner and ex-sheriff of Duval County, he got elected on his promise to create an "Empire of the Sun." The Everglades were just "wet desert," he said, useless to human beings, but if they were drained, "freedom, happiness, and prosperity" would inevitably follow.

"Americans did great things," one environmental historian wrote. "Therefore Americans would drain the Everglades." Broward himself lifted the first shovel of black muck. Dredgers and dynamiters went to work. When some objected that the Everglades should be studied before they were altered forever, he laughed: "I will be dead by that time. Let's get a few dredges and begin." Broward died in 1910 but the work continued, and by the mid-1940s his vision had been realized: There were 440 miles of canals, 47 miles of levees, 16 locks, and an assortment of weirs and drainage ditches that shunted water into the Atlantic. Lake Okeechobee—not much deeper than a fence post but 730 miles around—was hemmed in by high dikes. And 1,000 square miles of wetland had been transformed into grazing land for cattle and sugar cane fields.

Meanwhile, a growing chorus of voices tried to make itself heard over the din of development, people who saw another kind of value in the marshes and worried that more might be happening than was immediately apparent. It seemed too late to save the northern part of the Everglades, but they began to agitate for a sanctuary to be created out of the sprawling, still-untouched saw grass marshes at the state's southern tip. In 1947, they prevailed. Congress created Everglades National Park—the largest national park in the 48 contiguous states—a watery expanse of saw grass, which had been given the wonderful name "El Laguno del Espiritu Sanctu" (the Lake of the Holy Spirit) by early Spanish explorers.

We now know that the Everglades is a complex system of wetlands, unique in scope, delicacy, and richness of life. Across the flat Florida limestone tablerock stretch cypress swamps, pinelands, hardwood hammocks, and freshwater prairies ending in some of the world's finest mangrove swamps circling Florida Bay. For thousands of years, water inched southward, down the ever-so-slight gradient from Lake Okeechobee, in a shallow sheet flow, spreading across the Everglades to evaporate and fall as rain over much of Florida. The groundwater from the cavernous limestone aquifer that lies underneath the Everglades was and remains the primary source of fresh drinking water for southern Florida.

There was little understanding in the early part of the twentieth century of what we now call an ecosystem, and no one had any idea of how the Everglades affected the land around them and vice versa, or even where they began and ended. In 1947, Marjory Stoneman Douglas, a veteran journalist whose father had been among Governor Broward's most stubborn opponents, published a book that revealed its singular nature. *The Everglades: River of Grass* was beautifully written, historically exact, and remarkably prescient about Everglades' hydrology. Douglas was a tiny woman, although everything else about her seemed oversized—her courage, high spirits, and her eventual impact on South

Florida. When she first considered writing a book about the Everglades for a popular series on rivers, a hydrologist told her the Everglades weren't swamps at all but a subtle flow of water. She mused about definitions: "If it's running water and it comes curving down from Lake Okeechobee toward the Ten Thousand Islands, and if there are ridges on either side, maybe the ridges are an east bank and a west bank, and maybe the Ten Thousand Islands are a delta and maybe this really is a river." If her book changed the definition of the Everglades forever, the Everglades, in return, would transform Marjory Stoneman Douglas.

When she first saw the Everglades, great flocks of ibis, roseate spoonbills, egrets, storks, heron, and cranes were so thick that they seemed to darken the sun as they lifted over saw grass and water. Panthers, bears, alligators, deer, manatees, and hundreds of species of fish all flourished here. She was bewitched. "Nothing anywhere else is like them," she wrote of the wetlands, "their vast glittering openness, wider than the enormous visible round of the horizon, the racing free saltiness and sweetness of their massive winds, under the dazzling blue heights of space." Marjory Douglas was not a scientist, but she struggled hard to understand how it all worked. She argued that although the Everglades began at Lake Okeechobee, the system actually started farther north, in lakes near Orlando. From there, she reasoned, waters pushed south into the Kissimmee River and Lake Okeechobee, and then flowed slowly into three million acres of marsh, finally entering Florida Bay. Primarily, she saw that the Everglades began and ended with water.

Ironically, in 1947—the same year that the National Park was created— the Everglades suffered a grievous blow. Floodwaters from Lake Okeechobee, driven by back-to-back hurricanes, devastated property in the Everglades agricultural area north of the park and as far south as Miami Springs. Congress responded immediately: to protect farmland and the population of southern Florida, it ordered the Army Corps of Engineers to bring the waters under control and link its drainage systems.

The engineers would be at it for the next thirty years—furiously rerouting, channeling, lifting, draining, and shunting waters north and east of the park into an additional 1,400 miles of canals and levees and an interlocking system of spillways, weirs, dams, and locks. They built 181 primary control structures and 2,000 smaller ones. In a day, the Corps' pumping stations routinely pump 1.7 billion gallons of water into the ocean but in that same twenty-four hours they can pull as much as 20 billion gallons of water off the land. The Corps foreshortened the winding hundred-mile-long Kissimmee River by 50 miles, and turned it into a dead-straight ditch of water called, without a breath of remorse, Canal C-38. It facilitated navigation but cut the flow of water into Lake Okeechobee. An equally straight highway, the Tamiami Trail, cut across the Everglades from Miami east to Fort Myers, severing the park almost entirely from its sustaining sheet-flows of water. A levee on the park's eastern side, meant to protect urban areas from flooding, separated it even more radically from outside water. Free-flowing water was a thing of the past. Virtually all the Everglades water outside the park was locked up by the South Florida Water Management District, responsible for water flow; every move made by the organization to solve problems seemed to create even more intricate difficulties in the Everglades system.

With each successive ditch or control maneuver, the Everglades lost a little more life. The largest park in the United States became the most menaced. The water of Canal C-38, deprived of oxygen because it was too narrow for the wind to keep it aerated and too deep for thorough mixing, stopped moving and became stagnant. Lake Okeechobee, choked by runoff from cattle and sugar farms, began to die.[189] Fires swept the drying wetlands.

Fifty percent of the basin has been developed and much of the system is gone for good. And although Everglades National Park includes 1.5 million acres—20 percent of the original system—all of it is jeopardized since it is wholly dependent on water coming from the outside, most of it fouled by agriculture or development. Damage hasn't stopped at land's end. In shallow Florida Bay, mangroves and turtle grass are withering, temperature and salt levels are rising (the bay is saltier than the sea), and all kinds of marine life are diminishing, from wading birds to spiny lobsters. Algae alone seems to flourish here, turning the once brilliant blue bay a dull gray-green.[190]

Tourists, who come to the Everglades from all over the world, can still be dazzled by green expanse viewed from park overlooks. But there isn't much fish, bird, or animal life left in those reaches, and what little remains clusters around the remaining bits of water, or else nests on the unprotected watery northern side of the Tamiami Trail. Since the park's creation, bird populations have been reduced by ninety-three percent.

Even Napoleon Broward might be surprised to learn that the diggers and dredgers he dispatched with such enthusiasm have slowed the rain. Less rain falls in central and southern Florida now because the sheets of water that once covered and then evaporated off the limestone tablerock aren't there to be lifted by the sun. Since that rain is the ultimate source of all water in southern Florida, and because so much water is pumped into the Atlantic, the growing populations of Broward and Dade Counties have discovered that water supplies from the Biscayne Aquifer have been vastly diminished. Furthermore, the lowered water levels in the aquifer make the groundwater vulnerable to saltwater intrusion. The quandary can be identified simply: South Florida needs water, a lot of clean water. Beyond that fact, nothing is simple. But the dessication of the Everglades has become the problem of every South Florida citizen who wants clean water to drink, and many of those citizens have taken on its cause.

From the first, the fight to save the Everglades has been fierce. Guy Bradley, a warden hired by the National Audubon Society to protect egret rookeries against plume hunters in 1905, the same year that Napoleon Broward started digging, was found with a bullet in his neck. Marjory Stoneman Douglas's father, Frank Stoneman, a lawyer and publisher who had editorialized against the canals, was denied a judgeship by the governor for opposing his plans. Marjory Douglas herself came into the fray in the late sixties. This was nearly two decades after she wrote her book, but she more than made up for lost time. Persuaded by friends already deeply involved in the battle to stop the wreckage of the Everglades, she responded, as a friend of hers commented, "like a fire horse to an alarm." She was in her late seventies and legally blind, but it didn't seem to matter. Wearing her rakish trademark hats, she began traveling around the state, attending meetings, mobilizing supporters, calling press conferences, quoting Virgil or Boswell to make her points, and confounding those threatening the "Glades."

Journalist Steve Yates describes her appearance at a meeting of the Dade County Planning Department to discuss a vote on zoning in the east Everglades: "[T]he crowd parted, and a small figure in a large hat was led down to the microphone. 'Butterfly chaser!' one man yelled in what I took to be his worst insult. 'Go home, granny! You don't own any land here.' Another yelled, 'Mind your own business!' After spending a moment adjusting the microphone, Marjory Stoneman Douglas, then ninety-one, turned to the jeering mob. 'I can't see you back there,' she said in a clear voice. 'But if you're standing up, you might as well sit down. I've got all night, and I'm used to the heat.' It took, as she would say, the starch right out of them."

"They're all good souls," she said later. "It's just that they shouldn't be out there." Marjory Douglas never stopped believing that the Everglades could be saved. "Natural places, ecosystems, are not fragile," she wrote. "They are in the main tough as an old tire. The capacity of the earth for compensation and forgiveness after repeated abuses has kept the planet alive."[191] She too proved tough as an old tire. In 1970, Marjory Douglas, at the age of eighty, created the Friends of the Everglades. Often called "Marjory's Army," the organization started with a few hundred people and eventually swelled to more than five thousand. In collaboration with fifty other groups—among them the Izaak Walton League, the Audubon Society, and the Sierra Club—Douglas and her allies mastered the intricacies of Everglades hydrology and forged alliances with local community groups, like the Coral Gables Women's Club, Kiwanis clubs, and condominium associations—to fight for the protection of water well fields in the Biscayne Aquifer. State and local politicians began to respond. In one groundbreaking decision, land use and industrial zoning became the focal point for protecting an underground water supply. The solutions and legislation were not foolproof, but because citizens became involved, important precedents were made part of Florida's progressive water laws.

Douglas and her coalition stopped the building of an airport in the Everglades—twice. They prodded the federal government to mount a suit against Florida that would make it enforce its own regulations. They aroused the media and moved public opinion to a point where most Floridians agree that the Everglades must be saved. As a result of the work of these good, informed citizens, there are now massive plans underway to reverse Florida history, to undo the complicated system of canals and levees that the Army Corps put in.

The work has begun at the top of the state and will help turn the Kissimmee River, which at the hands of the Army Corps of Engineers had become Canal C-38, back into a free-flowing river. Since the Kissimmee is the source of South Florida's water systems, there can never be enough water in the system unless the Kissimmee is replenished. Part of the river has already been restored by Florida's efforts to buy up floodplain, reflood wetlands, remove locks, open oxbows, backfill twenty-two miles of C-38, and add eleven miles of new channel to old riverbed. Canal water has been diverted into bends and meanders once choked with vegetation and devoid of living creatures that are now clear and alive with wading birds and fish. Although engineering on the Kissimmee will cost upward of hundreds of millions of dollars, most Floridians seem to agree that it is feasible as well as desirable to restore the river to its predevelopment state.[192]

The restoration has brought together old foes, the environmental movement and the canal's perpetrators, the Army Corps of Engineers, once one of the most reviled engineering outfits anywhere. In 1969, Supreme Court Justice William O. Douglas charged that the corps had "no ecological standards. It operates as an engineer—digging, filling, and damming the waterways. And when it finishes, America the Beautiful is doomed." Known best for its dispensation of congressional pork, it had heedlessly drained, dredged, and filled wetlands all across eastern America. Seeing the impact of their works across a spectrum of landscapes, its engineers ought to have understood more quickly than anyone else the damage they were doing. Yet they have been slow to admit to any of it. Environmentalists, with little technical knowledge, were way ahead of them. "It is only proper that [the Army Corps of Engineers] be held accountable for its past errors in this field," writes engineer Samuel Florman. "But in a nation where people persist in living on cliffs which are crumbling into the sea, and build houses atop major earthquake faults, there is little likelihood that even the most farseeing engineers could have prevented the rush onto the low-lying plains. It is a lot easier to hold back torrents of water than it is to stand in the way of land-hungry Americans."

The Army Corps of Engineers is an institutional organ of Congress, and it is here that canals begin or end. "Engineering works are a technical answer to a problem at the local level," says Roland Clement, who served for several years on an Army Corps environmental committee designed to work toward reform. "The problem is first articulated by the business community—farmers or developers—and then it moves to the political level," he told me in his Connecticut home. "The Chamber of Commerce says to the congressman, 'Look, you've got to bring home the bacon. We need this and we expect you to push it through the legislature.' And then—because these men have had a lot of experience pushing through legislation—the appropriation bill is passed, and then later on the chief of the Army Corps must allocate the appropriation over its districts. Conservation groups object to what the Corps does, but don't challenge the process. The Army Corps was willing to change but only at the pace Congress would allow. If we don't recognize the institutions that constrain people's behavior—if we don't address that level of decision-making, or we attack individuals—instead of saying it's the institutional framework that must be changed, all we do is allocate blame. We neglect the realities of how business is done."[193]

Joe Podgor puts it another way. "The Army Corps of Engineers is like the skin on a lizard," he says. "The lizard is the government, and if the lizard turns right, the skin goes right with it." When I first met Podgor in 1992, he had no sympathy for the Army Corps. "I don't think they can do restoration. I don't think they have the talent," he told me. But, having watched them hard at work in Florida for years now, he feels differently. "At the local level, a handful of engineers were trying very hard to do their job. If developers didn't get something they wanted, they'd go over the heads of local engineers, all the way to Washington if necessary. The engineers got fed up with being cast as villains, embarrassed by newspaper stories."

Not only will the Army Corps restore the Kissimmee River, it will also plug, refill, and dechannelize the sprawling public works in southern and central Florida in order to restore as much as possible of the natural Everglades. They will breach canals, build gates, and elevate the Tamiami Trail to allow water to flow in a wide, slow spread that replicates sheet flow into the Everglades. "They are looking for success in the Everglades," says Podgor, "particularly a cadre of younger people who've gone through an educational system that teaches environmental management and ecology. They are trying to do this right."

There is no time to lose. The Everglades remains a system out of order. Fixing that system will be immensely complicated. Because it evolved over ages in response to an intricate system of fluctuating water levels, merely dumping water into it doesn't put the water where it needs to go, at the right depths or times to make the marshes come alive again. After well-meant attempts to deliver more water, scientists have watched grimly as wildlife populations continued to shrink. Since Congress declared a minimum water-delivery schedule in 1971, an ultimately important move, destructive surges have submerged nesting bird and alligator colonies, drowned deer, or flooded at the wrong times so that birds nested late and then, signaled by drying swamps that it was time to leave, abandoned nests with baby chicks. And while carefully placed and timed water flows are crucial to the Everglades, the quality of the water it receives is equally important. Microbiologists believe it to be the most sensitive sequence of wetlands anywhere in the world, and it is breaking down because the water farmers in the Everglades agricultural area are taking water out of the regional supply and returning it heavy with fertilizers, manure, and pesticides. Even though the levels of these contaminants are not high by waste-treatment standards, nutrient runoff is deadly in this hypersensitive network. Cattails and other non-native species fed by phosphorus from fertilizers aren't conducive to nesting by Everglades wildlife and crowd out saw grass and oxygen-creating periphytic algae, the base of the Everglades food chain.

Plans to address the polluted water that flows out of Lake Okeechobee are among the most difficult part of the Everglades Scheme because they directly affect the agricultural industry. North of the Everglades and south of Lake Okeechobee sits the sugar industry and its swaths of dark, peat earth farmland that stretch across South Florida, punctuated by white plumes of smoke rising out of refinery smokestacks. Sugar, a multibillion-dollar business—although it employs only around 16,000 farm workers, modest numbers far outweighed by the tourist industry—enjoys strong government support and an iron grip on Florida water politics.[194] After years of haggling and decisions out of Jacksonville, Miami, and Washington, re-engineering is still pending, and the biggest impediment to clean water is still sugar.

When the state of Florida wanted to add a penny to each bag of sugar to pay for the $760 million needed to clean up phosphorus in marshes, a long, bitter, $6 million court battle ensued, in which the federal government sued the South Florida Water Management District to enforce the District's own clean-water standards. The sugar industry pumped about $25 million into television and lobbying, the tax was defeated, and Florida's taxpayers were saddled with big-sugar's expenses yet again.[195] Ultimately, the cost to sugar growers to help clean up their own mess was $230 million. The citizens picked up the bill for the remaining $530 million. "The fact remains that growing sugar has been a disaster for the Everglades," *The Economist* reported in 1997. "Sugar operations are actually expanding there, despite a $1.5 billion federal program to buy back Everglades land used for sugar production. Without the web of sugar supports, none of this would have happened."

The federal government has removed some subsidies, and sugar farmers have been asked to clean up a portion of their discharges on their farms. Some of the farms say that they have cleaned up more than required. Although reductions in phosphorus, the most damaging fertilizer nutrient, have been achieved to something like fifty parts of phosphorus per billion gallons of water, the very smallest acceptable level is ten parts of phosphorus per billion parts water.

In 1999, an unexpected challenger to the polluters turned up, the sovereign Miccosukee Indian Nation, who live on 264,000 acres in a water-conservation area north of the Everglades Park and east of the Big Cypress wetlands. The Miccosukees were outraged at the pollution pouring out of farms and nearby cities into their wetlands. "The Everglades is our mother," pronounced tribal chieftain Billy Cypress. "And she is dying." Water being pumped into the Miccosukee marshes by the South Florida Management District at this moment contains between 70 and 150 parts of phosphorus per billion. "We're being used right now as a filter for everybody else's pollution," said Gene Duncan, director of Miccosukee water resources. "They are going to have to clean the water up to meet our standards." The tribe, led by their chief counsel Dexter Lehtinen—the heroic U.S. attorney who first filed suit against the state of Florida on behalf of the federal government—demonstrated to the Environmental Protection Agency that 10 parts of phosphorus per billion was a valid limit.

The Environmental Protection Agency approved the standard, forever changing the rules of pollution control in South Florida. "Miccosukees' Clout Shaping Glades Restoration," read the headline in the *Miami Herald*. "They did what nobody else had the guts to do," Joe Podgor trumpeted. "That water-quality standard can be used by the tribe to force polluters to clean up. It sets a standard for the rest of the Everglades." The South Florida Water Management District has since scrambled for new cleanup measures to meet the standard.

The struggle continues. Everglades Park hydrologist Bob Johnson told me that in the years that he and his colleagues have spent fighting legal battles for more and cleaner water, he's often felt that they've taken several steps back for every step forward. "But if we wait until the ultimate solution comes around, the Everglades will be lost," said Johnson. "That applies to both water quantity and quality. We don't see the ultimate solution in terms of deliveries to Everglades National Park. We don't even have the science completely down. But we know we can make incremental steps to get us closer and closer. We know enough about the system to know what needs to be done. We are moving in the right direction, but because the agricultural industry in South Florida really controls water management much more than urban or environmental sectors do, we don't have the political will right now for the ultimate solution."[196]

Dozens of scientists in the Everglades have worked hard at understanding the historical Everglades—how much water there once was, how it moved, where, when, and at what pace. The engineering work now being accomplished in conjunction with what the scientists have grasped about water management and ecosystems can save something important there, a remnant of the former Everglades, even if some things have been lost forever. But it will cost a lot, and as always, the final answer to what will happen here lies in the hands of politicians and those who elect them.

When I first visited the Everglades in 1991, the environmental coalition was fighting hard to make local and national politicians do their job. Revisiting eight years later, I was astonished to find that the battles had not let up, although players and sides had shifted here and there. Half of the land in Shark Slough has been purchased for the National Park, but the other half will be harder to acquire since much of it is already devel-

oped. The costs, including Kissimmee restoration, reengineering, filtration swamps, and land purchases, have escalated by about a billion dollars each year I've been away. "Everyone will say, 'Yes, we support
saving the Everglades,'" Bob Johnson said to me finally. "Until it means
money out of their pocket or reduced profits, land losses, or constraints
on growth. As soon as you pose any of those things, the process slows
down. In the meantime the Everglades are being trashed."

The fight for the soul of the Everglades has meant that its scientists
have become warriors too, working alongside whoever is working on
behalf of the park—the coalition of environmentalists, the Army
Corps, or even a prodded federal government. Bob Johnson, a working
hydrologist who would prefer to spend all of his time trying to solve the
physical problems of the Everglades, spends at least two days a week and
sometimes weeks at a time at outside meetings, wielding his science on
behalf of the park. Thanks to the work of Johnson and his scientific colleagues, the Everglades ecosystem is better and more thoroughly understood than it has ever been. Everyone involved in the issues of water
management in South Florida, a conglomeration of often-disparate private and public groups, is struggling to use that information to solve the
problems of the damaged system. "It's important for scientists and environmental groups and the public to be linked," says Johnson. "It's never
really happened before. But it seems to be happening in the Everglades.
It's exciting because the science you are doing doesn't just sit on the
shelf. It means something."

"It's not too late, or we wouldn't be working," Douglas remarked in a
1990 interview in the *Miami Herald* on her one-hundredth birthday.[197]
"We simply cannot let everything be destroyed. We can't do that, not if
we want water. We've got to take care of what we have. You can't stop
the pressures of population, but you must protect natural resources.
People have to be informed. They have to be taught about the nature of
the country, which is fragile. It means a constant fight."

The Everglades Coalition, in preparing a 1992 plan for twenty-first-
century regional water management, concluded: "It is ours to provide

the example and we have no choice but to do it. Here, now, where the future of the Everglades begins, we have the capacity to do it right."

"We don't have much experience in creating big engineering projects," I was told by the late Jim Webb, then head of the Wilderness Society's branch in Florida, in 1992. "We're 'glistening headwaters people.' It's a big, expensive undertaking. And it demands a large-scale structural fix. You don't get big public works done without a vigorous local sponsor. So you need to align forces." Webb was full of hope. He pointed out that there are many reasons to expect success in the Everglades restoration. "There is the underlying creation and expansion of the park," he said. "And it still rains. That raw resource is local and undamaged. We will be able to restore lost storage and connections. And there is a clear alignment of state and federal policies to regulate restoration. Here you have seen a genuine federal effort, which has not happened anywhere else.

"What is good for the Everglades is good for the general economy," said Webb finally. "For all of our problems, it's the most advanced case of ecosystem restoration anywhere."

I TRIED BUT FAILED to see Marjory Douglas in the winter of 1992. The first President George Bush had paid a recent visit to the Everglades and said some encouraging things that he promptly forgot about when he went back to Washington, where he proposed a rule that would implement a more developer-friendly definition of wetlands. Furious, Marjory Douglas held a press conference on her front lawn, holding up a sign saying, THE EVERGLADES ARE ABOUT TO BE BUSH-WHACKED! Then 101 years old, she came down with pneumonia the next day and was put in the hospital. I worried aloud to Joe Podgor about her age and frailty. Joe, who had been fighting alongside her for two decades, just grinned. "You have to remember that ever since I've known Marjory, she's been an old lady." The gallant lady died seven years later at the age of 108. In 1999, while I was again talking with Podgor about the Everglades battles, he spoke of his long experience with Ms. Douglas: "I expect politicians to be like Marjory Douglas. If she said something, she meant it, it was true, and she'd do it!"[198]

"The Everglades wars," wrote Marjory Stoneman Douglas in 1987, "are not over." They are not over yet, but in the Everglades we've seen that part of the answer to understanding and saving the Everglades rested with a tiny, determined old lady, wearing a hat pulled stylishly over one eye, who did her best work after the age of eighty. Douglas marshaled an incredible force of fellow combatants and simply never stopped fighting. Nathaniel Reed, former U.S. Under Secretary of the Interior, has said, "Because of her, what was considered to be heretical, far-out, impossible, even stupid, is now considered important, necessary, and doable."

In Florida, scientists, an aroused public, and government agencies are together trying to do something important. Although the outcome relies on the continued concern of the public—ordinary people like Marjory Douglas and the rest of us—the nature of public works is being reconsidered here. The utter ambitiousness of the attempt to restore what is left of the Everglades is breathtaking in its scope. If we can manage it, it will have real meaning for our future on the planet.

At the end of *The Everglades: River of Grass*, Marjory Stoneman Douglas spoke of the balance between the forces of life and death at work in the Everglades. She then wrote: "There is a balance in man also, one which has set against his greed and his inertia and his foolishness; his courage, his will, his ability slowly and painfully to learn, and to work together." In that ability to learn and cooperate lies our hope for the world's waters—and ourselves.

Notes

1 *New York Times,* August 15, 1993.

2 This is about twice as fast as the loss before 1980. Charles Harris of Cardiff University has found that ground temperatures in European mountain ranges have increased by up to 2° Celsius over the last hundred years. In Amsterdam in July of 2001, 1,800 climate scientists from 100 countries met to discuss the mounting evidence of man's involvement in global warming and warned that IPCC's predictions of climate models could be conservative.

3 In the past 25 years, the welling up of deep water in the eastern Pacific Ocean has decreased, along with its subsequent cooling effect on the ocean and creation of trade winds. This in turn is believed to mean more and stronger El Niño events.

4 The Abbey of Mareindal, the Commandery of the Teutonic Knights at Nes, the Fraternity of St. John the Evangelist, the Monks of Advard, and the monastic house at Witlenwierum were just some of the religious houses that drained peat bogs and built dikes.

5 An old Frisian manuscript contains a farmer's yearly instructions from an A-sega, that is the predecessor to a dikemaster: "Cleanse the sluices, clear the ditches, repair the highroads and home roads, heighten and strengthen the sea-burghs and dams, and make drainings underneath the roads that the water may pass; in springtime work at the sea-walls, in summertime cleanse the drainings from vegetation and throw this on the sides, and during midsummer work the whole long day in the fields."

6 According to a genealogical painting, the survivor was named Beatrix, who grew up, married, and had three children.

7 Andries Vierlingh as quoted in J. Van Veen, *Dredge, Drain, Reclaim: The Art of a Nation;* Martinus Nijhoff, Den Haag, 1962.

8 When the British decided to drain the Fenlands in the 17th century, they hired Dutch waterman Cornelius Vermuyden.

9 He even "created" a piece of land that is named after him, the Van Dixhoorn Triangle.

10 Construction Engineer Willem Zwigtman told me about dumping concrete. "When we made closures, we dumped blocks along the whole length. We did some experiments with helicopters that we got from the U.S. Army. They took a few concrete blocks and dumped them, thinking that the work may go faster. But when they dropped the blocks, the helicopter jumped up—way up. I remember one man from work that flew with those guys. He was sick as hell."

11 The Rijkswaterstaat awarded the contract to a consortium of seven companies: the Bal-

last Nedam Groep, Bos Kalis Westminster Groep, Baggermaatschappij Breejenbout, Hollandsche Beton Groep, Van Oord Utrecht, Kon. Volker Stevin Groep, and Aann. Comb. Zinkwerken. They were collectively called Dosbouw.

12 The *Mytilus* and other construction vessels, *Cardium, Ostrea,* and *Macoma* were designed and built by the Dutch firm BVS, Bureau voor Scheepsbouw Ir. P.H. de Groot b.v. They were all named after creatures of the estuary.

13 The barrier is closed an average of once or twice a year.

14 The safety record on the project was remarkable. "They started in '78 and I only know of one casualty," says Zwigtman. "If there were two, I think that's it," he told me. "Very good, very safe."

15 Flooding carries with it new concerns, since Holland's river water is loaded with PCBs, chlorides, and heavy metals washed in from Germany, Switzerland, and France. Holland's own industrial waste, manures, and fertilizers from intensive agriculture dirty waters too. The Dutch government, generally innovative in Green thinking, has ordered its farmers to cut their use of pesticides in half. Beatrix, Queen of Holland, addressed the problems in a 1988 speech: "The earth is slowly dying," she said. "We human beings ourselves have become a threat to our planet."

16 Eighty percent of the pollution is from the farms. Each year over 11,000 tons of nitrogen and 13,000 tons of phosphorus enter the lagoon. Over 500 tons of algae are pulled out of the lagoon daily in summer.

17 Schemes to reduce oil traffic here are being implemented gradually, so that industry can absorb the costs without going under. The production of some oil products will be shifted to other Adriatic ports.

18 Gentilomo has reservations concerning the effectiveness of pollution-absorbing plants: "My impression is that you may be simply transferring the problem to some other place. But right now this is the most practical solution for 80 percent of the pollution. You can manage industrial pollution and sewage by conveying polluted waters to treatment plants via pipes but agricultural pollutants are so widely distributed that you cannot push them inside a pipe.

19 The gates are installed on reinforced concrete foundations of rock, placed inside an excavated trench, linked to the foundation through connectors. Inside the foundations there are networks of pipes, water, electricity, compressed air, and, of course, tunnels for maintenance workers. Waterjets will clean debris from the gates and operating mechanisms. When the gates are up they will sit at a 45-degree angle.

20 The total length of channel openings to be gated is a third of a mile at Chioggia, a third of a mile at Malamocco, and half a mile at the Lido.

21 London's Thames Barrier Gates also sit on sills on the floor of the seabed and also rise up against flooding. It's a relatively small amount of cement to protect Venice.

22 Venice relies on septic tanks. "In the past, varying water levels encouraged oxygenation in septic tanks. Now due to the lowering of the city, oxygenation is reduced. In Venice we do not have room for inclined pipes which function by gravity, but simple and inexpensive American vacuum systems with small horizontal pipes are available, which move

waste matter by electric motors and compressors. To a Venetian, this is considered artificial and unacceptable." M. Gentilomo.

23 In an area of 40,000 acres irrigated by the upper Bari Doab Canal, subject to rising water tables, there were 12,000 deaths in 1908 alone. "One may justly arrive at the opinion that ague has not been introduced as a new thing into the canal-irrigated country, but that its area and period of prevalence and its intensity of attack have very greatly increased since irrigation from the canals was introduced," reported one Dr. Planck.

24 Although more than a million died in the drought and famine of 1837–38, crops on newly irrigated land were saved and many lives with them.

25 While word of the canal's opening in 1854 reached around the globe, local villages did not find it plausible that the holy Ganges waters could be moved. "The indignation and astonishment of the natives on the occasion is most graphic," wrote an attending missionary. " 'What!' said their spokesman. 'Do you dream that our great and holy goddess, Ganga, the mother of gods, is going to change her course at the bidding of you impious English?' " But the canal defied all its doubters. Its channels held 17 dams, 16 falls to ease water down steep slopes, 31 locks and navigable channels, and 10 miles of bathing ghats. Its designer, Proby Thomas Cautley, was knighted. Years later, William Willcocks, who spent much of his young life along the Ganges Canal, acquired three horses, whom he named Proby, Thomas, and Cautley.

26 Cromer, an able administrator, came from a well-known British banking family. Nicknamed "OverBaring," he was described by General Gordon as having "a pretentious, grand, patronizing way about him."

27 When, at the end of 1889, the canal clearance corvée was ended, Willcocks was ecstatic: "Within the first five years of the British occupation of the country we had swept away an institution which weighed heavily on every poor man in Egypt and which was at least 7,000 years old. . . . It was a great personal triumph."

28 He remained unshaken, even in the face of a telegram such as one he received prior to an oncoming deluge: "The engineer in charge of the Aft Section of the Damietta branch reports that the Nile has eaten away about 100 metres in length of the bank, and is continuing its work of destruction; he has telegraphed to Cairo for three million cwt. of stone and has himself, by order of his doctor, left his post and gone home, as he is very ill."

29 An incensed Cromer had discovered that the engineer was writing a comic history of Egypt under the English and forced Willcocks to sign a promise not to write anything while in government service. Willcocks, with four children to feed, agreed. There was further unpleasantness when Willcocks loudly suggested that (British) tennis players in Egypt should all be subject to the same rules, including Lord Cromer, who bent them in his own favor. A final affront came when Willcocks criticized a superior officer who was a friend of Cromer's, and when he received a sharp reply, sent a letter back suggesting that Lord Cromer and his chief "labored under the delusion that they could twist nature's tail as they did the tails of their subordinates."

30 Later, on Willcocks's Hindia Barrage in Persia, the river was closed by felling towers of

giant masonry cubes. Gunpowder was inserted into holes in some of the cubes and deto-
nated; the pillars collapsed into the breach.

31 The Pan American Canal was predated by a plan to cut a canal through Nicaragua.

32 In 1847, Brigham Young set his followers to diverting streams for farms in the valley of
the Great Salt Lake. The Mormons, who were from the East, had no experience of irrigation and worked by trial and error. They dug canals and ditches to carry water by gravity flow to farm plots, mostly without the use of surveying instruments. They made
levels from wooden pipes and water-filled glass bottles. One bishop leveled his canal
using a milk pan filled with water. They attached water rights to the land and organized
communal methods of water-sharing and responsibility for ditches. They proved that the
desert could "bloom like a rose," and in 15 years' time had 150,000 acres under irrigation, the first large-scale irrigation in North America. Mormon colonies spread from
Canada to Mexico, using their newly discovered methods of farming dry land.

Around that same time, gold was discovered in California at Sutter's Mill, and thousands of would-be miners rushed west. The miners invented hydraulic mining, a means
of processing great amounts of earth fast. "Hydraulicking" utilized rockfill dams and networks of flumes, descending in long tiers down mountainsides for the sluicing of gold.
Miners channeled fast-falling water into nozzles and turned them in streams onto mountainsides, washing whole hillsides down into ground sluices. While hydraulicking ravaged the landscape, farmers adopted their abandoned sluices and techniques.

33 Municipalities and private irrigation and power companies liked the thin and hollow
shapes because they could save as much as 60 percent in material.

34 Benjamin Holt, owner of Holt Manufacturing in California, invented the caterpillar tractor just before World War I. Steam shovels had been used to build the Panama Canal
around the turn of the century, but their size and lack of mobility meant that they were
not suited to most dam sites.

35 Chief design engineer on the Hoover Dam was John L. Savage.

36 As quoted in Hoover Dam, Joseph Stevens, University of Oklahoma Press, Narma, 1988.

37 The Hoover Dam helped win World War II, as its electricity-powered steel and aluminum mills made possible the building of one-fifth of America's fighter planes.

38 Six Companies' men sank the piers for the Golden Gate Bridge, and worked on the
Minidoka Dam, Cabinet Gorge in Idaho, the Taylor Park Dam, Box Canyon Dam, Davis
Dam on the Colorado, Missouri's Table Rock Dam, Parker, Ruby, Hungry Horse and
Chief Joseph dams, the Delaware Aqueduct Tunnel. They worked on part of the contract for Grand Coulee Dam, also on the Columbia River, which at 10.5 million cubic
yards of material would be the single largest concrete structure in the world. There
were construction booms on this project as high as a football field is long. Grand Coulee
was the first structure in the world to exceed the volume of the Great Pyramid of
Cheops. They called it a "man-made mountain range," and on completion it produced 40
percent of all energy used in the Northwest.

39 Pacific Bridge built docks for the Navy at Pearl Harbor. Bechtel built a naval base in the
Philippines, pipelines across the Canadian Rockies and in Alaska, oil refineries in

Curaçao, the West Indies, and the sub-Arctic, and ports, refineries, railroads, and pipelines across Saudi Arabia and the Arab Emirates.

40 Willcocks lost his own final water war along the Nile when he took after the British head of public works in Egypt, Sir Murdoch MacDonald, in an astonishing, some say viciously pursued, attack on MacDonald's plans for new dams in the Sudan, accusing the man of exaggerating figures and destroying hydrological evidence to justify cotton cultivation. Remembering Willcocks's confidence in his own beliefs—he once answered, when queried about his views of someone else's opinion, "I think too hard to think a thing I don't think"—one can be sure that the engineer led the charge driven by conviction. Nonetheless, the British Consular Court convicted Willcocks in 1921 of criminal slander and libel, although the old man was let off on good behavior.

41 Egypt became a constitutional monarchy in 1922. Britain, however, maintained several areas of jurisdiction (or reserve clauses) and kept troops in Egypt, especially around the Suez Canal.

42 Adrien Daninos proposed the High Dam in 1912 in order to generate hydroelectric power to industrialize Egypt.

43 The High Dam is not entirely to blame, since studies have shown that river water loses two-thirds of its sediment in 10,000 kilometers of canals north of Cairo. Sediment is also trapped upstream in Sudanese dams and irrigation systems on the Blue Nile, and downstream behind barrages on the Damietta and Rosetta branches in the delta.

44 Perennial irrigation often boosts schistosomiasis, since year-round canals are rarely drained or cleaned. The eggs are deposited by urine and fecal matter from humans and cattle into rivers and irrigation channels, where they are hosted and hatched in snails. The larvae then infect whatever hapless person or creature walks into the water. In a study of 1,800 people treated along the Nile, only 89 needed more than one dose of Praziquantel to be cured. Twenty-eight were not helped. China mounted a war against schistosomiasis in 1940, when at least 10 million people were infected, with an intense education program. The country treated infected people and created a simple water-seal toilet, which killed parasites in sedimentation tanks. The disease is no longer a serious threat there.

45 Wild vegetation around the shores of the reservoir has attracted both grazing wild animals and permanent settlements of nomads with cattle, whose migration routes were severed by Lake Nasser. Thousands of these former nomads, however, are infected with schistosomiasis.

46 The 1922 Colorado River Compact, an agreement based on flow records made during wet years, officially allots more water in a year than the river yields—16.5 million acre-feet of a river more likely to yield 14 million acre-feet. Without even accounting for evaporation, which removes another 2 million acre-feet, the Compact constitutes 2.5 million acre-feet of over-optimism. The water is delivered through adjuncts to the Hoover: the All-American Canal, which carries a full fifth of the Colorado's flow to California's Imperial Valley; the Colorado Aqueduct, which sends another billion gallons a day 240 miles to Southern California; the Central Utah Project, which runs to the Great

Salt Lake Basin; the Central Arizona Project stretching all the way to southern Arizona; and Canal Central and the Gila Gravity Main Canal, which move what's left over to Mexico. The states do not use all of their allotments, and since the lower-basin states have become used to gobbling up extra water, they are building preemptive projects just to make sure they don't lose their portions.

47 Wyoming farmer Beryl Churchill told me of a representative from Las Vegas who came to a Wyoming State water meeting looking to buy or lease some of their Colorado River water. " 'I've got my checkbook right here,' he said as he patted his back pocket," says Beryl. "Of course he got a cool reception. But the suggestion that Wyoming can name her price for water to be sold downstream is still around."

48 Las Vegas has been engaged in a variety of sensible things like paying Nevada farmers to let their land lie empty in order to save groundwater. Hotels like the Mirage reuse wastewater to create the displays so outrageous in a desert. A recent ordinance forbids developers to dig any more lakes, and encourages landscaping with desert plants, important since popular turf grass uses between 30 and 50 percent more water than any other green cover.

49 One-third of North American fish species exist in the West.

50 The Las Vegas water grabs have brought together people and organizations that fought each other in the past. Three Nevada counties and one in California have joined forces to fight the metropolis. Counties, towns, ranchers, farmers, the National Park Service, the U.S. Fish and Wildlife Service, Paiute Indians, and environmentalists have registered at least 3,600 formal protests with the state engineer since Las Vegas made its bid.

51 In the competition between cities and farmers, wildlife and wild lands—without monied interests to back them—are especially vulnerable. Reduced flows in rivers and deltas have trounced California's populations of Chinook salmon, delta smelt, and striped bass. In San Francisco Bay, the source of drinking water for 20 million people, overpumping has degraded the delta, devastated fish populations, and allowed its water to become saline. Mono Lake, a source of 15 percent of L.A.'s water, has been drained by 45 feet, its salt content doubled, decimating the food supply for migratory birds and creating a dangerous, potentially deadly dust hazard.

52 The action made it to the U.S. Supreme Court, which ruled in favor of the TVA.

53 There were from the very beginning severe differences between David Lilienthal and Board Chairman Arthur Morgan, especially about how to deal with the utilities. Morgan could not accept philosophical disagreements between himself and the other two board members and wrote articles in national magazines attacking them. Roosevelt, troubled by the dissension, eventually removed Arthur Morgan from the TVA board.

54 Coincidentally, private power company revenues rose because of increased use of electricity. In 1936 Commonwealth & Southern's net income rose $4 million over the preceding year. As the demand for power rose, new steam plants kept coal miners at work.

55 The new dams included two big ones, the Douglas Dam on French Broad Creek and Fontana on the Little Tennessee River. Both dams were near the facilities of the Aluminum Company of America (ALCOA) a longtime presence in the valley. Lilienthal

negotiated a brilliant partnership contract with ALCOA to buy the Fontana Dam site from them and operate six more of their dams within the TVA system.

56 All of South Asia's great rivers emanate from within short distances of each other in this part of the Himalayas; the Ganges, the Yamuna, the Indus, and the Brahmaputra. A major Indus tributary, the Sutlej pours out of Lake Manasarovar, south of Kailas, the "Lake of the Mind."

57 India and the Hindu people take their name from the Indus; the Sanskrit word *Sindhu* means "ocean." I understood better why the ancients, who had never seen the larger ocean, called the river by such a name, when my friend Mira spoke of crossing the Indus when she was a little girl. From the riverbanks, she said, she could not see the other side, just the frightening waters, stretching, it seemed, forever.

58 David Lilienthal, "Another 'Korea' in the Making," *Colliers,* August 4, 1951.

59 Various parties repeatedly suggested that Lilienthal arbitrate, but he always declined, judging at each point that his entry was not appropriate, yet privately holding out hope for an eventual role. Lilienthal, who remained involved behind the scenes throughout negotiations, was gratified to hear from the Chief Engineer of India, A. N. Khosla, of a measured conversation with Nehru, in which the Prime Minister said, "I would not deprive the common man of Pakistan of water any more than I would deprive the common man of India; on the other hand, we must not doom our own people to starvation for lack of water."

60 In April of 1964, the consortium gave another $315 million to Pakistan.

61 The treaty carefully avoided any position regarding Kashmir or Azad Kashmir, which Pakistan had occupied.

62 The bucket-wheel excavator was later taken to Egypt and the Sudan to build the Jonglei Canal.

63 Wheat and rice yields tripled in this thirty-year period. There were similar increases next door, in Haryana State.

64 Bhakra has had its share of problems. Water has been poured too freely on the farmland of the Punjab and there is poor management in the watershed where salinity and water-logging are taking a hard toll. The big dam was meant to have a productive life of 630 years, but because sediment is accumulating in a mound in the reservoir, it's thought that Bhakra's working life will be reduced by some 200 years. But it remains India's Hoover, doing its work and more.

65 This was in spite of the fact that demand for electricity had peaked and in some years even declined.

66 David Freeman knew David Lilienthal. "I was very fortunate," he told me. "I got to know Mr. Lilienthal in his later years personally, and well, and quite favorably. He was a remarkable man. He could take a mundane project and make it exciting. He had that talent. Not that TVA was mundane but some of the details were."

67 The Bonneville Power Administration helps make Freeman's case. It carries a $16.1 billion debt, much of it due to investments in nuclear power plants.

68 The Churchills use mostly surge irrigation, ditches, and gated pipe. It's still generally felt to be too expensive to put sprinklers in for a single yearly crop.

69 The Bighorn Basin, a prehistoric inland sea long ago gone dry, although between 4,000 and 5,000 feet above sea level, is some of the lowest land in Wyoming.

70 Wyoming state engineer Elwood Mead issued Bill Cody and his partner, Nate Salsbury, irrigation rights in 1897.

71 The Shoshone Project was the second federally built irrigation project in America; the first, only six months older, was on the Salt River in Arizona.

72 By 1910 the government had spent $3,360,000 on the dams, tunnels, canals, and ditches on the Shoshone Project, a bill that would be paid eventually by those who settled the land. The homesteaders received the land from the government under the Reclamation Act for a $10 filing fee, but they did have to pay the government's construction and drainage costs and ongoing operation and maintenance through an annual per-acre assessment.

73 The Buffalo Bill Dam irrigates roughly 100,000 acres. Two districts, the Garland and Heart Mountain divisions, share staff and equipment in an office, which delivers water to about 70,000 acres. Its young manager Ed Norlin started out working on an irrigation ditch dragline, and Watermaster Dean House is the son of a ditchrider. There are eleven ditchriders, now driving trucks instead of horses, who pick up orders, open headgates, and move water, each covering about 70 miles a day. Each farmer pays a water charge per acre to cover the district's costs and Buffalo Bill Dam and Reservoir operations. They do not pay for the water.

74 All four districts are now in the middle of another rehabilitation on the aging system. This one will cost $15 million and put covers on ditches, line canals, replace a leaking flume with a siphon, and make repairs on a dam and drops in the 83-year-old canal.

75 *Cadillac Desert,* Penguin, New York, 1986.

76 Even within the Shoshone Project there is divergence. Frannie Division has problems with bad soil and poor drainage. Much of that division was just plain misguided from the beginning. A map of Garland District shows a pattern of geometrically laid out lateral and sublateral canals that deliver water to a series of regularly spaced farms, but a map of Frannie offers a confused maze of long, expensive, curving canals servicing a few farms. Many of those living on Frannie have now taken jobs in town rather than keep trying to farm the difficult land. McCormick says that it's a waste of water and with what is known now about soil technology, Frannie would not be irrigated today. In 1903 Buffalo Bill predicted to his sister, "Frannie won't be anything hardly."

77 Approximately 90 percent of Pakistan's food comes from the Indus Basin.

78 The cities of Karachi, Hyderabad, and Sukkur are desperate for stable supplies of drinking water.

79 Pakistan's average fertility rate is 6.6 babies per woman, among the highest anywhere.

80 "A one-meter depth of irrigation (an amount normally applied in a single season) of even reasonable-quality water contains sufficient salt to salinize an initially salt-free soil." Daniel Hillel.

81 In its final reaches in Sindh Province, the plain slopes just two inches in a mile.

82 Pakistan's big dams are silting up and some of its smaller dams have become virtually ineffective. WAPDA conducted an Integrated Watershed Management program prior to

the building of Mangla Dam, which utilized foresters, civil engineers, and agronomists working with farmers to plant managed forests in key watersheds, to reclaim land for farming by using silt traps, and to check dams buttressed with vegetation. The Dam's Monitoring Organization reported a yearly reduction of 7,850 acre-feet of silt per year because of this program. Nothing like this was done at Tarbela or has been planned for other new dams, other than some reforestation.

83 A web of organizations under the giant umbrella of Pakistan's Water and Power Development Authority and the World Bank have developed a computer model of the Indus Basin, which includes most local aspects of water, soil, silt, and salt. They know, for example, where wheat is produced and where the cotton comes from, where bullocks are used and who has tractors. They know where waterlogging and salt are severe. Since 1960, Salinity Control Projects in Pakistan have put down thousands of kilometers of pipe drains.

84 Forty-eight percent of the land in the Sindh is salt-affected. Engineers are also thinking about building an outfall drain on the right side of the Indus river, more problematic because its bank is hillier.

85 The Left Bank Outfall Drain—under construction since the early 1960s—carries only a quarter of its intended loads, yet the government of the Sindh struggles to maintain the endless costs of pumping, repairing, and cleaning it.

86 Although all of the best dam sites in the country have been used, it seems certain that at least one more big one will go up. Kalabagh, at $11 billion, is the most likely dam. It will be built where the Indus enters the plains.

87 "In the author's opinion, most damage done by overgrazing of rangelands and salinization of irrigated lands is reversible, with exceptions that probably affect less than 2 percent of irrigated land." H. E. Dregne, 1990.

88 Salt-tolerant crops like samphire or salt bush (*Atriplex*), of which there are some 200 species, yield products from cooking oil to cattle feed.

89 There are new lining materials on the market, such as ironlike polymers that coat a canal's concrete, rendering it impermeable and extending its life. Remote monitoring, microwave-computerized water release, and interceptor canals at the ends of canal lines to prevent spillage all mean more crop to the drop. Drip irrigation, surge irrigation, low-energy precision application sprinklers that deliver water closer to the ground to reduce evaporation, and microsprayer, subsurface, and bubbler irrigation all save substantial amounts of water, and, while expensive, will pay their way in a hurry when water becomes scarce.

90 Disease has increased with large irrigation projects in developing countries, since standing water makes for happy snails and mosquitoes. If health officials and water bureaucrats and engineers work together to plan systems and work with local people, many problems can be averted. Channels to keep water moving or strategically timed lowering and raising of water levels behind dams and weirs are just two ways of preventing stagnant water and vector-borne disease.

91 David Seckler, *Water Scarcity in Developing Countries,* Winrock International Institute for Agricultural Development, Arlington, VA, 1993.

92   In Cameroon, large-scale private irrigation has been successful, while in Mali, large-scale public schemes have a miserable record. Small projects in Senegal and Cameroon have failed to attract farmers, but in the Niger, small, low-tech private schemes have done very well. In Rajapur, Nepal, the Dutch engineering firm Euroconsult helped farmers upgrade a century-old canal system that carried water to almost 22,000 acres. Euroconsult worked with farmers and engineers to solve the problems of a deteriorated irrigation system and inequitable water distribution.

93   USAID studies found that between 30 and 50 percent of diversions were lost in watercourses, which are operated communally. The World Bank identifies 35,600 watercourses in saline areas desperately needing renovation at about $600 million. Projected benefits are $235 million each year after completion, with an increase of about 4.6 MAF in water, at a cost two-thirds that of canal-lining. Traditional brick lining cuts water loss by 80 percent but doesn't last long. Rubberized and plastic lining materials, which are 100 percent effective, have been developed in the West, but Zuberi and his engineers still don't know how they will hold up under Pakistani circumstances: the sharp hooves of passing cattle, vandalism, the teeth of rodents, intense ultraviolet radiation, and high temperatures. And the materials are expensive.

94   The official name of the canal is the Indira Gandhi Canal, after India's former prime minister. In Rajasthan it is more often simply called the Rajasthan Canal.

95   A Maharaja of Bikaner, Ganga Singh, was the first to try to green this desert, when he built a canal 130 kilometers north to the Sutlej River in 1927. The notion of trying it on a larger scale arose just before India's independence, and took on political importance during negotiations between India and Pakistan over Indus waters, since it seemed to justify India's claim for more water.

96   Climate change, which resulted in the drying of the area, is thought to have occurred around 4000 B.C.

97   Urmul Trust helps 150,000 people in 300 villages make their land productive by organizing cooperatives and offering inexpensive land leveling with Urmul tractors. Out of a grant from the Canal Command Authority it pays much of the material cost for covering watercourses against dust storms, when farmers give labor. They help resolve disputes, manage distribution of water and maintenance of watercourses, and plant trees. Sanjoy Ghosh, its founder, was a hero of Gandhian proportions who disappeared in 1998 while fighting for the benefit of poor people in Assam.

98   There is much on paper that is worthy, such as reservation of 50 percent of the land for the landless, and the creation of agencies offering recourse for those needing information and help. But so much is only on paper. Concrete provision has been made, however, for penalties and responsibilities. The difficulties of dealing with government agencies— hard, long, and complicated at best—are even more forbidding here because there are three separate agencies managing the canal. The revenue department makes four.

99   One of the saddest stories in the Thar is of 16,000 families displaced by the Pong Dam in Himachal Pradesh, who were to be given land in the canal command in return for lost homes in the hills since it was water from Pong now being sent to Rajasthan. By 1991, only about a third of those eligible had been settled in the canal area, many of them with

terrible tales to tell of land grabbing, harassment, beatings, and even murder. The ous-
tees, India's word for project-displaced persons, also complained of few or no schools,
little or no health care, and no drinking water. Many were given uneven land—in need
of expensive land-leveling—or are served by faulty, dry canals. One farmer, Ganga
Ram, who lost a 25-acre green farm in Himachal, was in return given barren, unirri-
gated land in Rajasthan, but if he leaves it, he will be left with nothing at all. He is a
prisoner in the desert.

100 The construction of an outfall drain to dispose of saline drainage water from India's
land-locked Thar desert all the way to the sea, in order to make the canal even somewhat sustainable, would have cost $9 billion, essentially rendering it impossible to
build.

101 There are more birds here now, drawn to the seepage water along the canals, including
cranes and other species not native to this area. Some say the many new birds threaten
desert species, but those interactions are yet to be understood. In the seventies, a
Desert Wildlife Sanctuary was created but doesn't offer effective protection, since the
sanctuary itself, like everything in the Thar, is under pressure from outsiders.

102 Several rivers fall to the west out of these Alps, thousands of feet downward, onto the
plains. The Murray, the Darling, and Murrumbidgee, which run full into the center of
the continent, are lengthy but meager. Water in the lethargic Darling River, for example, takes a full two months to meander from its beginnings, 1,700 miles south, to a
meeting with the Murray. In drought years these rivers wither to a trickle of puddles.

103 In the 1890s, when arch dams were just beginning to go up in America, ten successful
arch dams already stood fast in Australia.

104 Committee founder William McKell, who was born and raised 50 miles from the
Snowy, went to America in 1945 where he was dazzled by the Tennessee Valley
Authority. In a report to the Australian Parliament, McKell quoted Franklin Roosevelt's words about a government corporation "possessed of the flexibility of a private
enterprise" to develop resources for "the general social and economic welfare of the
nation."

105 There were only a few aboriginal people then in the mountains, as most of them seem
to have been driven away or murdered long before.

106 Guthega Dam was 110 feet high and 456 feet long, built by Selmer Engineering of
Norway.

107 "Hoover's Jack Savage visited the Snowy as a consultant several times." Doug Price.

108 Erhardt Timmel, a member of the design team responsible for Germany's V-2 rockets,
became the head of physical sciences for the Snowy.

109 In the first issue of the *Snowy Mountains Magazine,* Hudson wrote of Sir Christopher
Wren's visit to the construction site of Saint Paul's Cathedral. When Wren asked three
workers what they were doing, the first two said they were dressing stone, "but the
third said, 'I am building a cathedral.' Let us all, from the pick and shovel men to the
top bosses, look at our work in the same way that the third man did, keeping before us
all the time that we are building the greatest work ever contemplated in this country
and one which the nation needs."

110 Tumut 1 power station is 244 meters below the earth. Tumut 2 is 366 meters under the mountains and accessed by a 1,000-meter tunnel spiraling into the ground (1,070).

111 The main Kaiser construction camp was called Sue City after Henry Kaiser's daughter-in-law.

112 Siobhan McHugh, *The Snowy: The People Behind the Power*, 1989.

113 The men used to say "a man a mile" in the tunnels, meaning that one man died for every mile of tunnel on the project. That's .06 man per mile, low in comparison with projects of this kind. Ninety-six men died at Hoover; and on the Mont Blanc tunnel, there were 1.4 deaths for every mile. Some feared that haste led to carelessness with human lives, but if the records are to be believed, the Snowy had a better safety record than many engineering projects. Some, such as Brad Collis in *The Snowy: The Making of Modern Australia,* have suggested that these figures are far too low.

114 The life after work for the 7,000 men who labored on the Snowy Scheme wasn't much to talk about. Prostitutes from Sydney enjoyed a booming business but didn't always stay long, as they were trailed by vice squads. Mostly there was drink. Seven or eight hundred men could be found in the bars on any night in Cooma. "Most men work two or three months, stash away a couple of hundred quid and clear off," commented a worker. "They can't stand the isolation. Nothing for them in the evenings, only grog and cards, no women."

115 Eucumbene, 381 feet high, was a full kilometer thick at the base.

116 Thiess Brothers, Humes Ltd., John Holland Construction, Monier, Tyree Electrical, and the Australian General Electric Company were among the big Australian companies on the Snowy Scheme. In 1967, the last major job, the Talbingo Dam, the highest dam in Australia, 530 feet up and 2,300 feet across, was awarded to Thiess Brothers.

117 Simply bolting rocks together won't work, since the whole bolted piece can—and has been known to—collapse. Snowy Mountain Associate Commissioner Tom Lang, who pushed its development, was called the Rockbolt King.

118 As quoted in "More Power to Australia," Jesse Ash Arndt, *Christian Science Monitor,* Oct. 14, 1959.

119 The Kaiser coalition built the Eucumbene-Tumut Tunnel, the big dam at Eucumbene, Tumut Pond Dam, Tumut 2 Dam, Khancoban Dam, the Tumut 2 Underground Power Station, and three large inlet and outlet tunnels to the Tumut 1 and 2 Power Stations. Morrison-Knudsen built a dam and power station at Blowering. Utah Construction built Tantangara Dam and Murrumbidgee-Eucumbene Tunnel, and with the help of Brown and Root, they built the Jindabyne and the Eucumbene-Snowy Tunnel.

120 As quoted in Siobhan McHugh's *The Snowy: The People Behind the Power,* 1989.

121 Quoted in *Scientific American*: "Nuclear Power," by James A. Lake and John F. Kotek. *The Economist,* March 28, 1998, on nuclear plants: "Not one, anywhere in the world, makes economic sense." Almost 220,000 tons of contaminated waste have been produced by 30 nuclear plants around the world since 1970 and are joined by an additional 10,000 tons a year. Recycling nuclear waste, which produces weapons-grade plutonium, also has a nasty by-product, pools of explosive water. It is also even more expensive than processing newly mined uranium. *Scientific American,* January 2002.

122 Natural gas is predominately methane, which is more potent than carbon dioxide but doesn't remain in the air as long.

123 In 1999, the EPA's Energy Star building programs saved 21 billion kilowatt-hours of energy and $1.6 billion in energy savings.

124 Wind as of yet is the only renewable source of energy, with installations that deliver upward of 5 megawatts. Wind energy is the cheapest of renewable technologies. It's also fast to install and bring online. When direct environmental costs are factored in, wind, geothermal, and biomass all have lower total costs than advanced (cleaner) coal. Geothermal energy and wind power are competitive with combined cycle plants. (DOE) "Theoretically, renewable energy can meet many times the world's demand." *Environment*, December, 2001.

125 At Talbingo, the largest dam, the Authority is slightly more aggressive and maintains seismic monitoring stations, since for two years after Talbingo's reservoir filled, there were enough small quakes to turn an occasional refrigerator sideways in the nearby town.

126 An exception is Tantangara, sitting in natural grasslands north of Eucumbene. In the summer, the wind blows grass into Tantangara's basin, where it decays, so there is natural hydrogen sulfide as well as nutrient loading from forests to the south, which means that Tantangara doesn't turn over by itself, and requires managing. The Snowy's annual cycle means there is some thermal stratification, which results in low amounts of dissolved oxygen, but the Authority is on top of it and monitors and manages storages for both water quality and longevity.

127 It has been suggested that the Snowy Mountains Authority is not financially viable, an argument for which a case can be made. Had the Scheme been built to simply carry Snowy Water down the Murray, the outcome would have been economically clearer. The tunnels and reservoirs that carry water into New South Wales increased capital costs by $49 million (Australian) annual operating costs by over $2 million, and costs per installed kilowatt by $44. A "Murray-only" Scheme would have been more profitable and more efficient. The two-river scheme was a political decision to appease the state of New South Wales, since without the agreement of both states, the Snowy Scheme could not have been built.

128 There are large reserves of brown coal or lignite in Australia, which, of poorer quality than black coal, has a high ash and water content.

129 When the Snowy was designed, the Department of Land and Water designated three percent flows at the bottom end of the river. Since there are still some tributaries on the lower Snowy below Jindabyne toward Orbost, the river does flow in the winter and spring.

130 The state governments and the Commonwealth will still own it. Unlike many government hydropower authorities around the world, the Snowy never became a bloated, self-propagating drain on public coffers, partially because it is owned jointly by states and has always been subject to debate.

131 A 1999 European Union directive opens up at least part of its member countries' electric markets. In America, utilities in about half of the 50 states have been opened to competition, with others moving in that direction.

132 Changing economics and politics are further complicated in the eyes of the Scheme's critics by chronic salt and waterlogging problems in irrigated areas of the Murray-Darling Basin. Although salt predates the Snowy—the ground of this former inland sea is loaded with it—extensive land clearing and irrigation have compounded a treacherous circumstance. It is argued sensibly enough, here as elsewhere, that if irrigators had to pay the real costs of their water, they might think more carefully about growing rice or cotton.

133 Fred Pearce, "Chill in the Air," *New Scientist,* May 1, 1996.

134 The idea of taxing carbon emissions to finance renewables has been around for a while as well as the sound proposal to stop subsidizing coal in any way. World Bank figures in the mid-nineties identified worldwide subsidies to coal-producing industry to be as high as $120 billion a year. Green energies from renewable sources, made available for slightly higher costs to those willing to pay for them, are not only popular but oversubscribed in Austria, England, Scandinavia, and in the Netherlands. Admittedly, this idea is limited to those with enough income to manage it, but it will be a combination of efforts that solve our energy problems.

135 In 1997 the United States produced less renewable energy than it did in 1987, even though around the turn of the century America has cut back on research into renewable energy.

136 The Mouths of the Ganges extend several hundred miles across the delta as the crow flies, from the Indian border to Chittagong. The measurement of ever-changing riverbanks in the delta, to my knowledge, has not been calculated.

137 The Farraka Barrage has long been a sore spot between India and Bangladesh, not to mention a bargaining point for India whenever they disagree with their neighbor. Treaties have recently been concluded between the two nations to regulate lower Ganges problems.

138 They suggest that to increase water in the dry season, Bangladesh use tube wells and pumping, and for floods they recommended a variety of interventions, ensuring that engineering projects include social, environmental, and economic impacts, better emergency preparation, a drastic improvement of warning systems, better science, and sharing information with India since Bangladesh's rivers come out of India.

139 In 1991 and in succeeding years, many lives were saved by storm-warning systems, and a remote sensing agency installed with the help of the World Meteorological Organization.

140 By 1812, New Orleans had built levees on the east bank of the river that began south of the city and finished 155 miles north. On the west bank, they stretched 1,000 miles.

141 The Flood Control Act of 1936 made floods a federal responsibility. The Flood Control Act of 1938 gave the government responsibility for all reservoir and channel modifications. Public Law 78-534 authorized upland treatment and flood damage control works in certain river basins. The Watershed Protections and Flood Prevention Act gave the Soil Conservation Service the go-ahead to work with the states on watershed management projects.

142 In 1993, the Corps claimed that their dikes saved more than was lost.

143 A congressional moratorium on the Flood Project for the whole valley due to an environmental question prevented the ACE from helping.

144 Boulder, Colorado, designed a recreational greenway along flood-prone Boulder Creek,

which also stores storm waters. In Ste. Genevieve, Missouri, town authorities got rid of risky properties and extended levees around the city center.

145 Robert J. Coontz Jr. says that in the Great Flood of 1993, rainfall measurements were hours old by the time the National Weather Service ran its models and that floodwaters hit Des Moines several hours earlier than hydrologic models had predicted. Some small, fast-moving storms slipped through the system entirely.

146 *Human Adjustment to Floods* (1945); *Optimal Flood Management: Retrospect and Prospect* (1966). From *Geography, Resources and Environment*, Vol. I, Selected Writings of Gilbert F. White, Kates & Burton, ed., 1986. Chicago: University of Chicago Press.

147 In 1938 Chiang Kai-shek used the river as a tremendous implement of war, dynamiting its dikes in order to stop the advance of Japanese armies with water. The Japanese were temporarily brought to a halt, but tens of thousands of Chinese died under the spreading waters.

148 *Wired, The Spirit of Mega*, July 1998.

149 In 1994, the World Bank approved $570 million for the 1,800-megawatt Xiaolangdi Project in China, of which one-fifth was for land and associated costs of 180,000 relocated people. Four foreign companies, including Italy's Impregilio and Lyonnaise des Eaux Dumez of France, built it along with Chinese engineers.

150 Over 100,000 people drowned along the Yangtze in 1931, and another 100,000 in 1935.

151 The World Bank, the Canadian Government, the Clinton Administration, and the U.S. Import-Export Bank adjudged the dam to be a bad idea, but engineering companies, heavy equipment producers, and banks were on queue to get the business. Three Gorges is now financed by Swiss, German, and Japanese money.

152 One of the most frightening seismic disasters anywhere happened near Pune in India at the Koyna Dam in 1962. Not long after the reservoir began filling, thousands of shocks were felt within 15 miles of the dam. Two big quakes, of magnitudes 5.0 and 5.5, hit in September 1967. Then in December, a 6.5 quake occurred, its epicenter precisely under the dam, which, although it cracked and seeped dramatically, did not give way. The dam-induced earthquake, however, killed 180 people. There were at least 30 quakes of magnitude 4.0 or higher around the dam for half a dozen more years.

153 In February of 1998, retired Chinese seismologist Huang Xiangning, who had been studying the relationship between clouds, droughts, and floods, warned that terrible floods would soon afflict southern China. Huang Xiangning told his government that they ought to mass-produce life preservers and inflatable boats, that they should move people, grain, medicine, and livestock to flood-safe areas. The government responded to Huang's warning, produced tents for the homeless, and strengthened dikes and reservoirs. Most important of all, they warned the public. Four months later, 4,150 souls were lost to the waters of the Yangtze River. The numbers, however horrific, were smaller than in years when there was no warning—such as the floods of 1991, when 7,300 died.

154 The *Jezirah* or *Gezira* means "island." The High Jezirah begins below the Anti-Taurus in Turkey. The lower Jezirah in Syria and Iraq is the limestone desert between the Tigris and Euphrates.

155 Turkey has offered an innovative if expensive appeasement, the "peace pipeline," a

3,000-kilometer-long double pipeline to carry water out of Turkey's Ceyhan and Sey-
han rivers to as far away as Saudi Arabia. At a cost of between $15 and $21 billion, some
countries claim that it is cheaper to desalinize sea water. Arab countries worry that
pipelines are open to sabotage and are hesitant to endow upstream states with political
leverage.

156 By noncooperation, Turkey forgoes outside funding for GAP.

157 Two new dams built on the Senegal and Baffing rivers between Senegal and Mauritania
inflated agricultural land values in the river basin and ignited violence. In the 1980s, the
elite Moors of Mauritania decided that they wanted the irrigable land for themselves,
and wanted all other landholders gone. Anthropologist Michael Horowitz, who worked
in the region for many years, told me the sorry tale: "We believe that the Mauritanians
rewrote their land legislation and then had a legal basis to get rid of the people. It was a
broad expulsion, not just the 70,000 herders and farmers but all Senegalese, including
Senegalese farmers with land rights on the Mauritanian side. But the critical population
were the 70,000 Pulaar and Fulme, and many were murdered and raped. The black
Mauritanians were exiled to Senegal where they remain as refugees."

158 T. Homer-Dixon and J. Blitt, *Ecoviolence: Links Among Environment, Population, and Security*, Rowman and Littlefield Publishers, Canham, MD, 1998.

159 Egypt has twice the gross national product of the Sudan and five times that of Ethiopia.
When the Technical Cooperation Committee for the Promotion of the Development
and Environmental Protection of the Nile met in Cairo in 1997, Egypt offered $100
million in participation for upstream states to develop the Nile.

160 The Sinai tunnels will irrigate 620,000 acres. Another 500,000 acres are to be irrigated
in the western deserts.

161 In years of heavy rain, the Sudd expands, appropriating seasonal grazing land from
farmers and nomadic grazers. Hundreds of thousands have died in the civil war in the
southern Sudan.

162 A second phase of the Jonglei Canal project involves storage in the equatorial lakes and
enlarging the canal.

163 War has meant that the Sudan, in spite of its own spiraling population, hasn't used
water ceded to it in the treaty of 1959. If the Sudan had been using its allowance, Lake
Nasser would have hit dead storage level in 1983 and the High Dam would not have
saved Egypt at all.

164 Water rights in the Western world have often centered around navigation. It wasn't
until 1973 that Colorado put in place a law that would guarantee water for fish and
wildlife, after a hundred years of first-come, first-serve.

165 The principles of *Shari'a* and also the influences of the French Colonial Codes de L'Eau
were incorporated in the 82 articles of the Ottoman *Majalla* pertaining to water, which
even now comprise a critical base for the laws of a number of Middle Eastern countries.
In all of these codes, the original right of water remains with the state.

166 Israel has reduced water use by about 25 percent through conservation. Its water scientists are some of the best in the world.

167 According to Peter Gleick in *Water in Crisis,* Jordan's yearly per capita water withdrawal is 173 cubic meters, while Israel's is 447 cubic meters. Syria withdraws 449.

168 Israeli hydrologist Dr. Miriam Lowi has written, in her book *Water and Power,* about the disparity in Arab and Jewish settlers' access to water: "While Palestinian Arabs have been prevented from sinking new wells for agriculture, Mekorot drilled thirty-six wells on the West Bank between July 1967 and 1989 for the domestic and irrigation needs of Jewish settlements. Of these, at least twenty are in the Jordan Valley and ten on the mountainous western fringe. And unlike Arab wells, which rarely exceed depths of 100 metres, those drilled by Mekorot [with superior technology] are between 200 and 750 metres deep. . . . The deeper the well (and the more geologically sound its location), the more abundant its water supply and the better equipped it is to resist contamination, salt water intrusion, and the harsh effects of drought. In addition, when two wells are located within the effective radius of each other, the deeper one tends to milk the water supply of the shallower one. When this is coupled with absence or sparseness of rainfall, the shallower well is sucked dry."

169 Roughly 30 percent of Israel's water comes from the Jordan, 40 percent from ground water, and 30 from treated wastewater, although figures differ wildly, as they do in regard to most Middle Eastern water sources. Israel claims historical usage of the water in the West Bank and moreover, since the aquifers pour westward, says that it belongs to them by right of natural flow. All the water sources are interconnected, and it is this that makes the hydropolitics so formidable.

170 Interview with writer Avi Shlaim.

171 "The '94 treaty included an interesting agreement to give 25 million cubic meters of Jordan's water to Israel, to freshen the water of Lake Tiberias and recharge the aquifer. It is to be given back to Jordan in the summer. The pipeline is built. They will also desalinate some of the springs they diverted in 1964. The major projects haven't been carried out yet because of a lack of funds." Sharif S. Elmusa. The agreement also opened the way for regional water development projects in the Jordan Valley and on the Yarmouk.

172 Arun P. Elhance, *Hydropolitics in the Third World, Conflict and Cooperation in International River Basins.* United States Institute of Peace Press, Washington, D.C., 1999. This is a superb book about water politics.

173 This, like many water figures, is a treacherous number. Some experts say 1,000; the figure 2,250 has been quoted by environmentalists; and Arundati Roy says that she was given the number 3,500 by an Indian agency. Using the internationally acknowledged figure of 50 meters as the criteria for a big dam, I have chosen 1,500 as likely.

174 Not far from Sardar Sarovar sits the 1972 Ukai Dam, with one of the world's largest spillways and a vast reservoir, built with World Bank help. The Bank's own 1990 study says that Ukai is silting up three times faster than was predicted and that its life span will be a pathetic one-third of what its builders had hoped. Fifty miles north, Sardar Sarovar will be three times as large.

175 Vincent J. Schaefer discovered in the mid-1940s that a tiny bit of dry ice produced mil-

lions of ice crystals when introduced into clouds of water droplets. His colleague, Dr. Bernard Vonnegut, improved the discovery with silver iodide.

176 The American company Atmospheric & Magnetics Technology sells a machine that, it claims, sucks moisture out of the air; then filters, purifies, and delivers it.

177 The city of Boston reduced water demand by 20 percent by fixing leaks in its distribution system, auditing industrial use and raising the cost of water. Grass is one of the most water-expensive plants on the planet, so in dry places by planting native species or drought-tolerant shrubs, which use 30 to 50 percent less water, we can make important savings.

178 A dripping faucet loses 36 gallons of water a day. Car-washes using recycled water shrivel use from 200 gallons to 15.

179 An award-winning project in Chile uses treated wastewater from El Teniente, the world's largest underground copper mine, to irrigate orchards and vegetable fields. Wastewater use on crops can be tricky and has been identified as the cause of some outbreaks of disease, so it's recommended for crops such as cotton. When used on food crops, it's best applied close to planting or well before harvest.

180 San Diego is especially vulnerable to drought, competition, and earthquake damage to the delivery system, and uses a process that combines a progression of treatment technologies in which wastewater passes through a water reclamation plant, then moves to an advanced water-purification facility, where it is treated by reverse osmosis to destroy viruses and parasites, then through ion exchange, low-pressure membrane filtration, and advanced oxidation, and after this well-monitored process is moved into the San Vicente Reservoir to be stored with other water. The reservoir water is treated again in the city filtration plant, so it's thoroughly purified before entering the water supply. The wastewater brine is pumped to the ocean. Some water out of the first treatment plant goes through a disinfectant process and is sent for nonpotable uses. A major part of San Diego's effort has been to thoroughly educate the public about the effectiveness of the process in order to overcome what they call "the yuck factor." The firm of Montgomery Watson worked with the city of San Diego to design the project.

181 A public information campaign and distribution of water-saving devices lowered use another 29 percent. In Sydney, politicians say that pricing water to cut demand ended calls for a new dam.

182 Charging water for agricultural use, which makes up well over 60 percent of all use, poses a far bigger problem than for domestic and industrial users. Right now worldwide, less than 5 percent of the investment costs in agricultural infrastructure are recovered.

183 In the worst drought in 50 years in the American West—2001 and 2002—some electric utilities paid farmers in Oregon and Idaho for the use of their water to generate hydropower for coastal cities. Those particular farmers have the right to sell their water for a year or even two, but if they sell it for several in a row, they can lose their claim to the water altogether. Naturally they are reluctant to do this.

184 The federal government can override the rights of farmers, as in southern Oregon where the government denied farmers their water in order to protect migrating salmon.

185 For a view of California water marketing, see Wade Graham, "A Hundred Rivers Run Through It," *Harper's*, June 1998.

186 The 1981 Water Code identified rights as being consumptive, nonconsumptive, permanent, or continuous, and available by outright purchase, bidding for surplus water, and by prescription or application. Consumptive uses allow the holder to use water without replacement; nonconsumptive means the holder can use it but must restore it. Permanent rights are unrestricted rights to unexhausted sources. Continuous rights allow water use 24 hours a day, every day, discontinuous rights specify periods, and alternate rights indicates successive rights divided among two or more users. Deeds specify use. Renato Gazmuri Schleyer, "Chile's Market-Oriented Water Policy: Institutional Aspects and Achievements," World Bank Technical Paper 249. Washington, D.C., 1999.

187 A local manufacturer turns them into Carp-Fert, a potent fish fertilizer.

188 The scale of the work is vast and cannot satisfactorily be chopped down into a few pages. In fact it occupies many books, an entire box of which was dispatched to me on request. Murray-Darling Basin Commission, GPO Box 409, Canberra, ACT 2601.

189 Even in the Everglades Agricultural Area, the oxidized drained peaty soil is subsiding and has begun to wear down to bedrock. Five feet of soil have been lost since the 1920s and it is estimated that it could all be gone within the next 20 years.

190 In the bay, 80,000 acres of sea grass bottom have already been denuded or are dying.

191 From the afterword of Marjory Stoneman Douglas's *The Everglades: River of Grass*, 1987 edition.

192 The Water Resources Development Act of 1992 authorized the Kissimmee River Restoration and the Headwaters Revitalization projects. Recreational and commercial benefits due to the restoration will be substantial—fishing, to name just one.

193 The Army Corps of Engineers, still in the process of changing, is more a fixer and restorer than it ever has been. It works on "nonstructural solutions," such as the Big Muddy National Fish and Wildlife Refuge, and undoing years of their own handiwork in the Everglades. It's not all rosy, since in 2000 the Pentagon released a report saying that three ACE engineers doctored facts in a Mississippi locks case in response to agribusiness interests and to hike up the ACE's own construction budget.

194 In one example, the SFWMD, in direct conflict with the U.S. Fish and Wildlife Service, took 28,000 acre-feet of water out of the Arthur Marshal Loxahatchee Wildlife Refuge (part of the Everglades system) at a time when the water levels were critically low, to irrigate crops in the EAA.

195 The lawsuit was on behalf of the U.S. Fish and Wildlife Service and the National Park Service.

196 Big sugar is not the only polluter damaging the Everglades. Other agriculture, developers, tourism, and the wildly growing population of South Florida also add to detrimental discharges.

197 Margaria Fichtner, *Miami Herald*, April 7, 1990.

198 Marjory Stoneman Douglas was one of a large number of people who have made the Everglades their cause and without whom the current progress would not have been made. Some key figures are Ernest Coe, Garald Parker, Art Mitchell, Jim Webb, Paul Parks, and Dexter Lehtinen, but any list omits many important people.

Glossary and Abbreviations

For those who don't understand building but would like to know more, I heartily recommend *The Construction Dictionary* published by the National Association of Women in Construction, P.O. Box 6142, Phoenix, AZ 85005.

Acre-foot: One acre-foot of water is 326,000 gallons or enough water to cover one acre, one foot deep. It is about enough to supply the normal needs of one or two families for a year.

Aquifer: Underground geological stratum that can hold and move water.

Alkali: Water-soluble salts of alkali metals, principally sodium and potassium. Especially prevalent in arid areas.

Alluvial Fan: Deposits of alluvium or sediment in a place where the gradient of a river slows, an earth delta.

Alluvium: Sediment or detritus carried by a river.

Archimedes Buoyancy: A principle defined by Archimedes, which states that the force which buoys a submerged mass is equal to the weight of fluid it displaces.

Blanket: An impervious layer underlying dams or dikes.

Barrage: Similar to a weir, a barrage is a low earthen or masonry dam built to raise water levels behind it high enough to divert water into canals, but a barrage is equipped with gates that lower and raise.

Braided Rivers: In river deltas, rivers sometimes cross and recross each other and are then called braided rivers.

Caisson: 1. A structure used in underwater work. In the case of dam building, a large hollow concrete structure, a base piece for a dam.

2. A wheeled vehicle for transporting artillery shells. In the Netherlands, actual caissons used in Allied assaults were used to close dikes.

Cavitation: In heavily rushing waters, unstable vacuums develop, which implode against walls, acting like chisel blows and digging holes that then cause the phenomenon to worsen.

Cement: Any of various mixtures of clay and limestone, usually combined with an aggregate to form concrete. It solidifies when water is added. Cement is the common name for Portland Cement, so called because of a resemblance to a quarried stone from Portland, England.

Check Dam: A very short concrete or embankment dam usually used to stop erosion.

Concrete: An artificial stonelike material formed by mixing aggregates—sand, gravel, shale, pebbles—and cement and water.

Cubic Meter: 219.97 gallons.

Cusec: Water flow in the amount of one cubic foot per second.

Diamond Drilling: The rotation of a diamond-set annular bit under pressure to cut into rock. The rock core is extracted from the drill hole in a core barrel. It's an expensive procedure.

Dike: A thick wall of earth or embankment, as much as ten times thicker at its base than at its top, that keeps water out or holds it in. Along the Mississippi, dikes are called levees, and in South Asia, bunds.

Dragline: A dragging rope. Also an excavating crane with a bucket or scoop, dropped from a boom by a cable and dragged to the crane base.

Feddan: Egyptian land measure. 1.038 acres.

Evapotranspiration: The loss of water from soil or plant or water surface into vapor.

Eutrophication: The smothering of a body of water which has become polluted and subsequently overgrown by algae or other aquatic plants, then depleted of oxygen.

Gelignite: Gel dynamite.

Gigawatt Hour: One million kilowatt hours.

Gigaliter: One thousand megaliters.

Gravity Dam: A dam that holds back water by its own weight.

GNP: Gross National Product. The total value of goods and services, usually produced in a year's time, and the primary indicator of a nation's wealth.

Grout: Thin, coarse mortar poured into various narrow cavities such as masonry joints or rock fissures. Grouting seals subterranean channels and any cracks in the structure of a dam.

Groyne or Groin: A jetty or waterfront spur dike, which extends into the water to protect against erosion. Also, a river training work that directs flow, traps sediment load, or establishes width of the river basin.

Head: In the production of hydropower, the distance water falls to generate electricity—either low, medium, or high.

Headworks: The regulatory works on a canal, consisting of a weir or barrage and head regulator or gated intake structure.

IUCN: International Union for the Conservation of Nature.

Kilowatt Hour (kWh): A standard unit of electrical energy; the amount of electricity dissipated by a one-kilowatt device in one hour. It refers to output and is the means by which electricity is sold.

Littoral: Seashore or intertidal area.

Low-head Hydro: Small-capacity, micro- or mini-hydroelectric plants, with a small drop or fall of water.

MAF: A million acre-feet.

Masonry: Construction of shaped units of stone, brick, cinder block, or concrete, usually bonded with mortar or cement.

Mastic: Adhesive or sealing compounds using bituminous materials.

Megaliter: One million liters. Two and a half megaliters will fill an Olympic-size swimming pool.

Meander: A portion of curving stream flow, whose length is more than one and a half times its downstream distance.

Megawatt: Electrical capacity. A million watts or one thousand kilowatts.

Monsoon: Wind systems that cause large temperature differences between land and ocean resulting in extensive periods of seasonal rainfall. South Asia receives most of its water in two monsoon periods, the northeast monsoons caused by winds off the Indian Ocean and the southwest monsoons blowing off the Arabian Sea.

Morphology: The study of the shape and form of individual organisms.

Mortar: A mixture of lime or cement, or a combination of both, with sand and water, used as a bonding between bricks, stones, etc.

NGO: A nongovernmental organization.

Osier: Any of a variety of willows. Often used as foundation for earthen waterworks.

Penstock: The pipeline or pressure tunnel that carries water from a dam's reservoir to turbines in the power plant.

Pile: A long, slim piece of steel, concrete, or timber that is put into the ground on its end to be used as a foundation or support.

Polder: A low-lying area of land, which has been artificially drained and is surrounded and protected by dikes.

Pore-pressure: Water pressure in saturated soil.

Pneumatic Drill: Drill operated by air or the pressure of exhaustion (compressed air).

Portland Cement: A kind of hydraulic cement made by burning a mixture of limestone and clay in a kiln. It is so called because of its similarities to a kind of clay from Portland, England.

Overtopping: When the height of water in a reservoir is higher than the dam and flows over it, it has overtopped the dam.

Radial Gate: A dam gate with a horizontal pivot (fixed) axis and curved water face.

Reinforced Concrete: Concrete reinforced with steel bars, which help it resist tensile stress.

Sheepsfoot Rollers: Rollers with protruding feet for tamping and compacting earth.

Spillway: The opening passage built into, through, or around a dam to discharge excess water if the reservoir is full. A stilling basin or apron at the bottom of the spillway dissipates the force of falling water.

Stilling Basin: Depressions at the foot of a dam or spillway, designed to reduce the energy of falling or flowing water.

Subsidence: Sinking of the ground surface.

Talus: A sloping mass of rocky fragments.

Technocracy: A government or social system controlled by technicians.

Technocrat: A technical expert, especially one in a managerial or administrative position. An adherent or proponent of technocracy.

Thakur: A Rajasthani nobleman.

Uplift: In dams, the force of water pressure pushing up from water that has found its way between the dam and its foundation.

WAPDA: Pakistan's Water and Power Development Authority.

Watt: A unit of power. One thousand watt hours is equal to one kilowatt hour.

Weir: A low masonry or earthen structure built to raise water levels so flow can be shunted into canal intakes. Usually built to be submerged in flooding.

Selected Bibliography by Subject

General Engineering and Water-Related Subjects

Books and papers

Adams, W.M., *Wasting the Rain: Rivers, People and Planning in Africa*. University of Minnesota Press, Minneapolis, 1992.

Allan, Tony, and Andrew Warren, *Deserts, the Encroaching Wilderness*. Oxford University Press, New York, 1993.

Anderson, Terry L., ed., *Water Rights, Scarce Resource Allocation, Bureaucracy, and the Environment*. Pacific Institute for Public Policy Research, San Francisco, 1993.

Berger, John J. ed. *Environmental Restoration: Science and Strategies for Restoring the Earth*, University of California, Berkeley, 1988.

Bottrall, Anthony, *Managing Large Irrigation Schemes: A Problem of Political Economy*. Overseas Development Institute, London, 1985.

Brown, Ellen P. and Robert Nooter, *Successful Small-Scale Irrigation in the Sahel*. World Bank Technical Paper 171, Washington, D.C., 1992.

Burton, Ian, Robert W. Kates, Gilbert F. White, *The Environment as Hazard*. Guilford Press, New York, 1993.

Caufield, Catherine, *Masters of Illusion: The World Bank and the Poverty of Nations*. Henry Holt, New York, 1996.

Clark, Robin, *Water: The International Crisis*, Earthscan Publications, London, 1991.

Cohen, Joel E., *How Many People Can the Earth Support?* W. W. Norton, New York, 1995.

Condit, Carl W., *American Building Art: The Nineteenth Century*. Oxford University Press, New York, 1960.

Desowitz, Robert S., *New Guinea Tapeworms and Jewish Grandmothers: Tales of Parasites and People*. Avon Books, New York, 1987.

Dorcy, Tom, ed., *Large Dams, Learning from the Past, Looking at the Future, Workshop Proceedings*. IUCN/World Bank, Gland, Switzerland; Cambridge, England; and Washington, D.C., April 11–12, 1997.

Dupriez, Hugues, and Philippe De Leener, *Ways of Water, Run-off, Irrigation and Drainage*. Macmillan Terres Et Vie, Belgium, 1990; London 1992.

Eisenberg, Evan, *The Ecology of Eden*. Knopf, New York, 1999.

Elhance, Arun P., *Hydro-Politics in the 3rd World, Conflict and Cooperation in International River Basins*. United States Institute of Peace, Washington, D.C., 1999.

Florman, Samuel C., *Blaming Technology*. St. Martin's Press, New York, 1982.

————, *The Civilized Engineer*, St. Martin's Press, New York, 1987.

————, *The Existential Pleasures of Engineering*, St. Martin's Press, New York, 1976.

Ghassemi, F., with A. J. Jakeman and H. A. Nix, *Salinisation of Land and Water Resources, Human Causes, Extent, Management & Case Studies*. Centre for Resource and Environmental Studies, The Australian National University, Canberra, 1995.

Gleick, Peter H., *Water in Crisis: A Guide to the World's Fresh Water Resources*. Oxford University Press, New York, 1993.

Gore, Albert. *Earth in the Balance: Ecology and the Human Spirit*, Houghton Mifflin, Boston, 2000.

Headrick, D. R., *The Tools of Empire: Technology and European Imperialism in the Nineteenth Century*, Oxford University Press, New York, 1988.

Hislop, Drummond, ed., *Energy Options: An Introduction to Small Scale Renewable Energy Technologies*, Intermediate Technology Publications, London, 1992.

Hillel, Daniel, *The Efficient Use of Water in Irrigation, Principles and Practices for Improving Irrigation in Arid and Semi-Arid Regions*, World Bank Technical Paper, Washington, D.C., 1987.

————, *Out of the Earth: Civilization and the Life of the Soil*. University of California Press, Berkeley, 1991.

Ingram, Helen, *Water Politics: Continuity and Change*. University of New Mexico Press, Albuquerque, 1990.

Jansen, Robert B., ed., *Advanced Dam Engineering for Design, Construction and Rehabilitation*. Van Nostrand Reinhold, New York, 1988.

Jansen, Robert B., *Dams and Public Safety*. U.S. Department of the Interior, Bureau of Reclamation, Boulder, CO, 1983.

Kates, Robert W., and Ian Burton, "Geography Resources and Environment," vol. I, *Selected Writings of Gilbert F. White;* vol. II, *Themes from the Work of Gilbert F. White,* University of Chicago Press, Chicago, 1986.

Kirmani, Syed S., and Rangeley, Robert, *International Inland Waters, Concepts for a More Active World Bank Role*, The World Bank, Washington, D.C., 1994.

Kollgaard, Eric B., ed., *Development of Dam Engineering in the United States*. Prepared in commemoration of the Sixteenth Congress of the International Commission on Large Dams, Pergamon Press, New York, 1988.

LeMoigne, Guy, William K. Easter, Walter J. Ochs, and Sandra Giltner, *Water Policy and Water Markets, Selected Papers from the World Bank's Ninth Annual Irrigation and Drainage Seminar*, World Bank Technical Paper Number 249, Washington, D.C., 1992.

Linsley, R. K., Joseph B. Franzini, David L. Freyberg, and George Tchobanoglous, *Water Resources Engineering*. McGraw-Hill, New York, 1992.

McCaffrey, Stephen C., "Water, Politics and International Law" in *Water in Crisis: A Guide to the World's Fresh Water Resources*, Peter H. Gleick, ed., Oxford University Press, New York, 1993.

McDonald, Adrian T., and David Kay, *Water Resources: Issues and Strategies*, Longman Scientific & Technical, Essex, England, 1988.

McKibben, Bill, *The End of Nature*, Random House, New York, 1989.

Organization for Economic Co-operation and Development, *Pricing of Water Services*, Paris, 1987.

Payne, Robert, *The Canal Builders: The Story of Canal Engineers Through the Ages*. Macmillan, New York, 1959.

Pearce, Fred, *The Dammed*. Bodley Head, London, 1992.

————, *Green Warriors: The People and the Politics Behind the Environmental Revolution*. Bodley Head, London, 1991.

Petroski, Henry, *Beyond Engineering*. St. Martin's Press, New York, 1987.

————, *To Engineer Is Human: The Role of Failure in Successful Design*, St. Martin's Press, New York, 1985.

Plusquellec, Hervé, Charles Burt, and Hans W. Wolter, *Modern Water Control in Irrigation*, World Bank Technical Paper, Washington, D.C., 1994.

Postel, Sandra. *Last Oasis: Facing Water Scarcity*, W.W. Norton, New York, 1992.

Powledge, Fred, *Water, The Nature, Uses, and Future of Our Most Precious and Abused Resource*, Farrar Straus Giroux, New York, 1982.

Ramage, Janet, *Energy: A Guidebook*. Oxford University Press, New York, 1999.

Rich, Bruce, *Mortgaging the Earth: The World Bank, Environmental Impoverishment, and the Crisis of Development*. Beacon Press, Boston, 1994.

Rogers, Peter, *America's Water: Federal Roles and Responsibilities*. Twentieth Century Fund Book, MIT Press, Cambridge, MA, 1993.

Safadi, Raed, and Hervé Plusquellec, *Research on Irrigation and Drainage Technologies, Fifteen Years of World Bank Experience*, World Bank Technical Paper, Washington, D.C., 1991.

Serageldin, Ismail, *Toward Sustainable Management of Water Resources,* World Bank Technical Paper, Washington, D.C., 1995.

Shady, Aly M. et al., *Management and Development of Major Rivers*. Oxford University Press, Calcutta, 1996.

Shapley, Deborah, *Promise and Power: The Life and Times of Robert McNamara*. Little, Brown, and Co., Boston, 1993.

Smith, Norman, *Man and Water*. Peter Davies Ltd., London, 1975.

————, *A History of Dams*. Peter Davies Ltd., London, 1974.

Van Slyke, Lyman P., *Yangtze: Nature, History and the River*. Addison-Wesley, New York, 1988.

World Bank, *Water Resources Management*, World Bank Policy Paper, Washington, D.C., 1993.

World Resources Institute, *1998–99 World Resources: A Guide to the Global Environment*, Oxford University Press, New York, 1988.

Articles

Calvin, William H., "The Great Climate Flip-flop," *The Atlantic Monthly*, January 1998.

Easterbrook, Greg, "Warming Up," *The New Republic*, November 8, 1999.

———, "Global Warming Meets the Prodigal Eagle," *The Economist*, October 11, 1997.

Gelbspan, Ross, "A Good Climate for Investment," *The Atlantic Monthly*, June 1998.

———, "The Heat Is On," *The Atlantic Monthly*, December 1995.

Graham, Wade, "A Hundred Rivers Run Through It: California Floats Its Future on a Market for Water," *Harper's*, June 1998.

Holoren John P., "Climate Challenge," *Environment*, June 2001.

Lewis, Michael, "Where Credit Is Due, The Funny Money of the World Bank," *New Republic*, February 13, 1995.

McKibben, Bill, "A Special Moment in History, The Future of Population," *The Atlantic Monthly*, May 1998.

Millich, Lenard and Robert G. Varady, *Managing Transboundary Resources, Environment*, Boulder, CO, October 1998.

Parlange, Mary, "Eco-nomics," *New Scientist*, February 6, 1999.

Pearce, Fred, "Deserts on Our Doorstep," *New Scientist*, July 6, 1996.

Pearce, Fred, "High and Dry in the Global Greenhouse," *New Scientist*, November 10, 1990.

Pearce, Fred, "Not Warming, but Cooling," *New Scientist*, July 1994.

Pearce, Fred, "Quick Change," *New Scientist*, November 14, 1998.

Richardson, Tom, and Rhodes Trussell, "Taking the Plunge," *Civil Engineering*, September 1997.

Welch, Ross M., Gerald F. Combs Jr., and John M. Duxbury, "Toward a Greener Revolution," *Issues in Science and Technology*, Fall 1997.

Williams, Phillip B., "Flood Control vs. Flood Management," *Civil Engineering*, May 1994.

Australia

Aplin, Graeme, et al. *Global Environmental Crises: An Australian Perspective*. Oxford University Press, South Melbourne, New South Wales, 1995.

Bolton, Geoffrey, *Spoils and Spoilers: Australians Make Their Environment, 1788–1980*. Allan & Unwin, Sydney, 1981.

Collis, Brad, *Snowy: The Making of Modern Australia*. Hodder & Stoughton, Rydalmere, New South Wales, 1990.

Crab, Peter, *Murray-Darling Basin Resources*. The Murray Darling Basin Commission, Canberra, 1997.

Dovers, Stephen, ed., *Australian Environmental History*, Oxford University Press, South Melbourne, New South Wales, 1994.

Hancock, W. K., *Discovering Monaro: A Study of Man's Impact on His Environment*. Cambridge University Press, Cambridge, England 1972.

Huxley, Elspeth, *Their Shiny Eldorado*. William Morrow, New York, 1960.

McHugh, Siobhan, *The Snowy: The People Behind the Power*. W. Heineman, Port Melbourne, Victoria, 1989.

Murray-Darling Basin Commission, *Blooming Algae*. Video cassette, Canberra, 1996.

————, *Salt Trends, Historic Trends in Salt Concentration and Saltload of Stream Flow in the Murray-Darling Drainage Division*, Canberra, 1997.

————, *Salt*. Video cassette, Canberra, 1996.

Pilger, John, *A Secret Country: The Hidden Australia*. Knopf, New York, 1991.

Powell, Joseph M., *The Emergence of Bioregionalism in the Murray-Darling Basin*. Murray-Darling Basin Commission, Canberra, 1993.

Seddon, George, *Searching for the Snowy: An Environmental History*. Allen & Unwin, St. Leonards, New South Wales, 1994.

Snowy Mountains Hydroelectric Authority, *Engineering Features of the Snowy Mountains Scheme*. Snowy Mountains Electric Authority, Cooma, New South Wales, 1972.

Wigmore, Lionel, *Struggle for the Snowy: the Background of the Snowy Mountains Scheme*. Oxford University Press, Melbourne, New South Wales, 1968.

Young, Ann R. M., *Environmental Change in Australia Since 1788*. Oxford University Press, South Melbourne, New South Wales, 1966.

China

Hangzhou Regional Centre (Asia Pacific) for Small Hydro. *Small Hydro Power in China: A Survey*, Intermediate Technology Publications, London, 1985.

Luk, Shiu-Hung, and Joseph Whitney, eds., *MegaProject: A Case Study of China's Three Gorges Project*. M. E. Sharpe, London, 1993.

Van Slyke, Lyman P., *Yangtze: Nature, History and the River*, Addison-Wesley, New York, 1988.

Winchester, Simon, *The River at the Center of the World*, Penguin Books, London, 1997.

Articles

Pringle, James, "Valley of the Dammed," *Times Magazine*, October 28, 1995.

Egypt and the Nile

Abate, Zewdie, "The Integrated Development of Nile Basin Waters," from *The Nile, Resource Evaluation, Resource Management, Hydropolitics and Legal Issues*. Centre of Near and Middle East Studies, London, 1990.

Abu-Zeid, Mahmoud, "Environmental Impact Assessment for the Aswan Dam," in Biswas, Asit and Qu Geping, eds., *Environmental Impact Assessment for Developing Countries*, Oxford University Press, Calcutta, 1983.

Ambrose, Stephen E., *Eisenhower, The President*. Simon and Schuster, New York, 1984.

Collins, Robert O., "Historical View of the Development of Nile Water," from *The Nile, Resource Evaluation, Resource Management, Hydropolitics and Legal Issues*. Centre of Near and Middle East Studies, London, 1990.

Collins, Robert O., "The Jonglei Canal: Past and Present of a Future," conference paper, The International Nile Conference, Tel Aviv University, May 19–22, 1997.

———, *The Waters of the Nile: Hydropolitics and the Jonglei Canal: 1900–1988*. Markus Weiner, New York, 1994.

Howell, P. P., and J. A. Allan, *The Nile, Resource Evaluation, Resource Management, Hydropolitics and Legal Issues*. Centre of Near and Middle East Studies, London, 1990.

Little, Tom, *High Dam at Aswan: The Subjugation of the Nile*. John Day Company, New York, 1970.

Moorehead, Alan, *The Blue Nile*. Harper & Row, New York, 1962.

———, *The White Nile*. Harper & Row, New York, 1960.

Nutting, Anthony, *Nasser*. Constable, London, 1972.

Phillips, John, *Kwame Nkrumah and the Future of Africa*. Frederick A. Praeger, New York, 1960.

Plusquellec, Hervé, "The Gezira Irrigation Scheme in Sudan," World Bank Technical Paper, Washington, D.C., 1990.

Waterbury, John, *Hydropolitics of the Nile Valley*. Syracuse University Press, Syracuse, 1979.

Willcocks, Sir William, *Egyptian Irrigation*. E. & F. N. Spon, London, 1904.

———, *From the Garden of Eden to the Crossing of the Jordan*. E. & F. N. Spon, London, 1919.

———, *The Irrigation of Mesopotamia*. E. & F. N. Spon, London, 1911.

———, *The Nile in 1904*. E. & F. N. Spon, London, 1913.

———, *The Nile Reservoir Dam at Assuan and After*. E. & F. N. Spon, London, 1901.

———, *Report on Perennial Irrigation and Flood Protection for Egypt*. National Printing Office, Cairo, 1894.

———, *Sixty Years in the East*. W. Blackwood & Sons, London, 1913.

———, *The Sudd Region of the White Nile and the Harnessing of Its Waters*. National Printing Department, Cairo, 1918.

Articles

White, Gilbert, "The Environmental Effects of the High Dam at Aswan," *Environment*, September 1988.

Europe

Felltham, Owen, *A Brief Character of the Low Countries Under the States*. A. Seice, London, 1661 (Duyckinck Collection, N.Y.P.L.).

Fitzmaurice, John, *Damming the Danube, Gabcikovo and Post-Communist Politics in Europe*. Westview, Boulder, Colorado, 1996.

Huisman, Pieter, *The Rhine, Artery of Europe: Human Interference in the Delta and International Cooperation in the Basin of the Rhine*. Ministry of Transport and Public Works, Den Haag, 1992.

Intergovernmental Panel on Climate Change, *Strategies for Adaptation to Sea Level Rise*, J. Gilbert (New Zealand) and P. Vellinga (Netherlands), Chairmen. Rijkswaterstaat, Den Haag, 1990.

Lambert, Audrey M., *The Making of the Dutch Landscape*. Academic Press, London, 1985.

Porter, Nancy, producer, *Will Venice Survive Its Rescue?* Produced for television by NOVA, WGBH, Boston, 1992.

Rijkswaterstaat & Dosbouw, *Delta Finale*, film, Den Haag, the Netherlands, 1990.

———, *Eastern Scheldt Storm Surge Barrier*, Den Haag, 1979. English edition of *Cement Magazine*, published by the Netherlands Concrete Society, December 1979.

Rijkswaterstaat, *Rising Waters: Impacts of the Greenhouse Effect for the Netherlands*, Ministry of Transport and Public Works, Tidal Waters Division, Den Haag, 1991.

———, *A New Coastal Defense Policy for the Netherlands*, Ministry of Transport and Public Works, Den Haag, 1990.

Schama, Simon, *The Embarrassment of Riches: An Interpretation of Dutch Culture in the Golden Age*. Knopf, New York, 1988.

Van Veen, J., *Dredge, Drain, Reclaim: The Art of a Nation*. Martinus Nijhoff, Den Haag, 1962.

Van De Ven, G. P., ed., *Man-made Lowlands, History of Water Management and Land Reclamation in the Netherlands*. International Commission on Irrigation and Drainage, Uitgeverij Matrijs, Den Haag, 1993.

Articles

Bandarin, F., "The Venice Project: A Challenge for Modern Engineering," *Civil Engineering*, November 1994.

Bernauer, Thomas, and Peter Moser, "Reducing Pollution of the River Rhine: The Influence of International Cooperation," *Journal of Environment and Development*, Thousand Oaks, California, December 1996.

"Out of the Mud, Some Lessons," *The Economist*, May 27, 1995.

"Wet, Wet, Wet," *The Economist*, April 28, 1998.

Pearce, Fred, "Dam Truths on the Danube," *New Scientist*, September 1994.

"Rising Water Drowns Opposition to Slovakia's Dam," *New Scientist*, July 16, 1999.

South Asia

Ahmad, Masood, and Gary P. Kutcher, *Irrigation Planning with Environmental Considerations: A Case Study of Pakistan's Indus Basin*. World Bank, 1992.

Ali, Imran, *The Punjab Under Imperialism*. Princeton University Press, Princeton, NJ, 1988.

Allchin, Bridget and Raymond. *The Rise of Civilization in India and Pakistan*. Cambridge University Press, Cambridge, England, 1993.

Asian Region Technical Department, *Flood Control in Bangladesh*, World Bank Technical Paper, Washington, D.C., 1990.

Azad, Maulana Abul Kalam, *India Wins Freedom*, Orient Longmans Ltd., Hyderabad, Andhra Pradesh, 1988.

Baber, Zaheer, *The Science of Empire: Scientific Knowledge, Civilization and Colonial Rule in India*. State University Press of New York, Albany, NY, 1996.

Baker, D.E.U., *Colonialism in an Indian Hinterland: The Central Provinces, 1820–1920*. Oxford University Press, New Delhi, 1993.

Berkoff, D.J.W., *Irrigation Management on the Indo-Gangetic Plain*, World Bank Technical Paper, Washington, D.C., 1990.

Bhakra Beas Management Board, *Bhakra Beas Projects*. Chandigarh, Punjab, 1985.

Byrnes, Kerry J., *Water Users Associations in World Bank Assisted-Irrigation Projects in Pakistan*, World Bank Technical Paper, Washington, D.C., 1992.

Canadian National Broadcasting Corporation, *The Dammed*, documentary film for *The Nature of Things*, Toronto, Ontario, 1995.

Doolette, John B., and William B. Magrath, eds., *Watershed Development in Asia*, World Bank Technical Paper, Washington, D.C., 1990.

Etienne, Gilbert, *Rural Change in South Asia*. Vikas Publishing House Pvt. Ltd., New Delhi, 1995.

Faeth, P. and R.P.S. Malik, *Agricultural Policy and Sustainability: Case Studies from India, Chile, the Philippines and the United States*, World Resources Institute, Washington, D.C., 1993.

Farmer, B.H., *Agricultural Colonization in India Since Independence*. Oxford University Press, London, 1974.

Feldhaus, Anne, *Water and Womanhood: Religious Meanings of Rivers in Maharashtra*. Oxford University Press, New York, 1995.

Fisher, William F., ed., *Toward Sustainable Development: Struggling Over India's Narmada River*. M. E. Sharpe, Armonk, New York, 1995.

Gadgil, Madhav, and Ramachandra Guha, *This Fissured Land: An Ecological History of India*. Oxford University Press, Delhi, 1992.

Ghosh, Sanjoy, and S. Ramanathan, *Irrigation Management Turnover: A Users Perspective: The Case of the Indira Gandhi Canal, Rajasthan*. Urmul Trust, New Delhi, 1995.

Gulhati, Niranjan, *Indus Waters Treaty*, Allied Publishers Pvt. Ltd., New Delhi, 1973.

Headrick, Daniel R., *Tentacles of Progress, Technology Transfer in the Age of Imperialism, 1850–1940*. Oxford University Press, New York, 1966.

Hodson, H. V., *The Great Divide: Britain-India-Pakistan*. Oxford University Press, Karachi, 1969, 1985.

Karim Koshteh, Dr. Mohammed Husain, *Greening the Desert, Agro-economic Impact of Indira Gandhi Canal of Rajasthan*. Renaissance Publishing House, Delhi, 1995.

Khan, Mohammed Ayub, *Friends Not Masters: A Political Autobiography*. Oxford University Press, New York, 1967.

Khan, Fazle Karim, *A Geography of Pakistan, Environment, People and Economy*. Oxford University Press, Karachi, 1991.

Kirmani, S. S., *Comprehensive Water Resources Management: A Prerequisite for Progress in Pakistan's Irrigated Agriculture*, Working Paper Prepared for Consultative Meeting, Washington, D.C., March 1991.

Kumar, Dharma, ed., *The Cambridge Economic History of India, volumes I and II*, Cambridge, England, 1983.

Lall, J. S., *The Himalaya, Aspects of Change*. Oxford-India Press, Delhi, 1995.

Lilienthal, David E., *The Journals of David Lilienthal, volumes I, III, and IV*, Harper & Row, New York, 1966.

Lipton, Michael, and John Toye, *Does Aid Work in India? A Country Study of the Impact of Official Development Assistance*. Routledge, London, 1990.

Maloney, Clarence, and K. V. Raju, *Managing Irrigation Together; Practice and Policy in India*. Sage Publications Pvt. Ltd., New Delhi, 1994.

Mason, Phillip, *The Men Who Ruled India*. W. W. Norton, London, 1985.

Mathur, Kuldeep, and Niraja G. Jayal, *Drought, Policy and Politics*. Sage Publications Pvt. Ltd., New Delhi, 1993.

Morse, Bradford, and Thomas R. Berger, *Sardar Sarovar, Report of the Independent Review*. Resource Futures International, Toronto, Ontario, 1992.

Menon, V. P., *The Transfer of Power in India*. Princeton University Press, Princeton, NJ, 1957.

Moraes, Frank, *Jawaharlal Nehru*. Macmillan Company, New York, 1956.

Michel, Aloys Arthur, *The Indus Rivers: A Study of the Effects of Partition*. Yale University Press, New Haven, CT, 1967.

Nehru, Jawaharlal, *The Discovery of India*. Anchor Books, Garden City, NJ, 1960.

————, *Toward Freedom: An Autobiography*. Beacon Press, Boston, 1961.

Pakistan National Committee of ICID, *Irrigation and Drainage Development in Pakistan*, Lahore, Pakistan, 1991.

Pakistan Institute for Environment-Development Action Research, *Increasing Irrigation Efficiencies Through Improved Community Management in Selected Chaks of Irrigated Punjab*, Islamabad, 1994.

Paranjpye, Vijay, *High Dams on the Narmada: Studies in Ecology and Sustainable Development*. Indian National Trust for Art and Cultural Heritage, New Delhi, 1990.

Qutub, Syed Ayub, *Environmental Impact of Agricultural Policies: A Case Study of the Linkages Between Irrigated Agriculture and Upland Watershed Protection*. Pakistan Institute for Environment-Development Action Research, Islamabad, 1993.

Rahman, Mushtaqur, *Land and Life in Sindh*. Ferozens Pvt. Ltd., Lahore, Pakistan, 1993.

Repetto, Robert, *The "Second India" Revisited: Population, Poverty, and Environmental Stress over Two Decades*. World Resources Institute, Washington, 1994.

Rogers, Peter, Peter Lydon, and David Seckler, *Eastern Waters Study: Strategies to Manage Flood and Drought in the Ganges-Brahmaputra Basin*. Irrigation Support Project for Asia and the Near East, USAID, Washington, D.C., 1989.

Sainath, P., *Everybody Loves a Good Drought: Stories from India's Poorest Districts*. Penguin Books India, New Delhi, 1996.

Saleth, R. Maria, *Water Institutions in India, Economics, Law and Policy*. Commonwealth Publishers, New Delhi, 1996.

Scudder, Thayer, *Supervisory Report on the Resettlement and Rehabilitation Component of the Sardar Sarovar Project*, World Bank, Washington, D.C., May 28, 1989.

Schafer, Howard B., *Chester Bowles, New Dealer in the Cold War*, Harvard University Press, Cambridge, MA, 1993.

Shah, Ashwin A., *A Technical Assessment of Two Non-Government Water Projects in Gujarat and Two Government Water Projects for Their Implications on Water Policy*. Unpublished paper, Scarsdale, New York, 1997.

Sheth, Pravin, *Narmada Project, Politics of Eco-Development*. Har-Anand Publications, New Delhi, 1994.

Singh, Kushwant, *A History of the Sikhs, Vol. 2*. Princeton University Press, Princeton, NJ, 1966.

Singh, Patwant, *The Sikhs*. Knopf, New York, 2000.

Spear, Percival, *A History of India, Vol. 2*. Penguin Books Ltd., Harmondsworth, Middlesex, England, 1965.

Stephens, Ian, *Pakistan*. Ernest Benn, London, 1964.

Stone, Ian, *Canal Irrigation in British India*. Cambridge University Press, Cambridge, England, 1984.

Thapar, Romila, *A History of India 1*, Penguin Books, Harmondsworth, Middlesex, England, 1966.

Thukral, Enakshi Ganguly, *Big Dams, Displaced People: Rivers of Sorrow, Rivers of Change*. Sage Publications New Delhi, 1992.

Truman, Harry S., *Memoirs*. Vol. 2, *Years of Trial and Hope*. Doubleday & Co., Garden City, NY, 1956.

Verghese, B. G., *Waters of Hope, Himalaya-Ganga Development and Co-operation for a Billion People*. Oxford and IBH Publishing Co., New Delhi, 1990.

————, *Winning the Future: From Bhakra to Narmada, Tehri, Rajasthan Canal*. Konark Publishers Pvt. Ltd., New Delhi, 1994.

Vohra, B. B., *Care of Natural Resources*. Indian National Trust for Art and Cultural Heritage, New Delhi, 1988.

————, *The Greening of India*. Tata Energy Research Institute, New Delhi, 1984.

————, *Land and Water Management Problems in India*. Ministry of Home Affairs, New Delhi, 1982.

————, *Managing India's Water Resources*. Indian National Trust for Art and Cultural Heritage, New Delhi, 1990.

Ward, Andrew, *Our Bones Are Scattered: The Cawnpore Massacres and the India Mutiny of 1857*. Henry Holt, New York, 1996.

Wescoat, James L. Jr., and Robin M. Leichenko, *Complex River Basin Management in a Changing Global Climate: The Indus River Basin in Pakistan, A National Assessment*. University of Colorado, Boulder, 1992.

West Pakistan Water and Power Development Authority, *Barrages and LinkCanals of the Indus Basin Project*. WAPDA, Karachi, 1967.

Whitcombe, Elizabeth, *Agrarian Conditions in North India, 1860–1900*. University of California Press, Berkeley, 1972.

Willcocks, Sir William, *Lectures on the Ancient System of Irrigation in Bengal and Its Applications to Modern Problems*. University of Calcutta, Calcutta, 1930.

————, *The Restoration of the Ancient Irrigation of Bengal*. University of Calcutta, Calcutta, 1894.

————, *Sixty Years in the East*. W. Blackwood & Sons, London, 1913.

Articles

Rahmani, Asad R., "Just Deserts, the Story of the Indira Gandhi Nahar Project," *Sanctuary Magazine*, vol. XIV, no. 1, Bombay, 1994.

Turkey and the Middle East

Biswas, Asit K., ed., *International Waters of the Middle East: From Euphrates-Tigris to the Nile*. Oxford University Press, Bombay, 1994.

Biswas, Asit K., J. Kolars, M. Murakami, J. Waterbury, Aaron Wolfe, eds., *Core and Periphery: A Comprehensive Apporach to Middle Eastern Water*. Oxford University Press, Bombay, 1997.

Chalabi, Hasan Dr., and Tarek Majzoub, "La Turquie, les eaux de l'Euphrate et le Droit Internationale Publique," *Water in the Middle East, Legal, Political, and Commercial Implications,* a conference, Centre of Near and Middle East Studies, London, and the Centre of Islamic and Middle Eastern Law, SOAS, University of London, November 1992.

Cooley, John K., "Middle East Water: Power for Peace," *Middle East Policy,* vol. 1, no. 2, 1992. Middle East Policy Council, Washington, D.C.

Du Bois, François, "Regulating the Competitive Use of Fresh Water Resources: Compensation and Exploitation," *Water in the Middle East, Legal, Political, and Commercial Implications*, a conference, Centre of Near and Middle East Studies, London, and the Centre of Islamic and Middle Eastern Law, SOAS, University of London, November 1992.

Elmusa, Sharif S., *Negotiating Water: Israel and the Palestinians*. Institute for Palestine Studies, Washington, D.C., 1996.

————, *Water Conflict, Economics, Politics, Law and the Palestinian-Israeli Water Resources*. Institute for Palestine Studies, Washington, D.C., 1997.

————, "The Water Issue and the Palestinian-Israeli Conflict," Information Paper no. 2, Center for Policy Analysis on Palestine, Washington, D.C., 1993.

Hillel, Daniel, *Rivers of Eden: The Struggle for Water and the Quest for Peace in the Middle East*. Oxford University Press, New York, 1994.

Kolars, John F., and William A. Mitchell, *The Euphrates River and the Southeast Anatolia Development Project*. Southern Illinois University Press, Carbondale, IL, 1991.

Lowi, Miriam R., *Water and Power, The Politics of a Scarce Resource in the Jordan River Basin*. Cambridge University Press, Cambridge, England, 1993.

Mallat, Dr. Chibli, *The Rights That Attach to Water, Customs and the Shari'a——A Legacy of Principles and Institutions. Water in the Middle East, Legal, Political, and Commercial Implications*, a conference, Centre of Near and Middle East Studies, London, and the Centre of Islamic and Middle Eastern Law, SOAS, University of London, November 1992.

Mehling, Marianne, *Turkey: A Phaidon Cultural Guide*. Phaidon Press Ltd., London, 1989.

Naff, Thomas, and Matson R. C., *Water in the Middle East, Conflict or Cooperation?* Westview Press, Boulder, CO, 1984.

Ponting, Clive, *A Green History of the World*. Sinclair Stevenson, Ltd., London, 1991.

Rogers, Peter, and Peter Lydon, *Water in the Arab World, Perspectives and Prognoses*. Harvard University Press, Cambridge, MA, 1994.

Tamrat, Imeru, "The Nile—A Consideration of the Issues," *Water in the Middle East, Legal, Political, and Commercial Implications*, a conference, Centre of Near and Middle East Studies, London, and the Centre of Islamic and Middle Eastern Law, SOAS, University of London, November 1992.

Willcocks, Sir William, *The Restoration of the Ancient Irrigation Works on the Tigris, or, The Re-creation of Chaldea*. National Printing Department, Cairo, 1903.

Woolley, C. Leonard, *The Sumerians*. Norton Library, New York, 1965.

Articles

Geffen, Harry, "Holy Water, Raising the Dead Sea," *The Toronto Globe and Mail*, Toronto, May 14, 1994.

Huband, Mark, "Politics Bottles Up Water Crisis," *Financial Times*, October 15, 1998.

Kolars, John, "Trickle of Hope," *The Sciences*, November/December 1992.

Ponting, Clive, "Historical Perspectives on Sustainable Development," *Environment*, vol. 32, no. 9, 1990.

Starr, Joyce R., "Water Wars," *Foreign Policy*, Spring 1991.

"Water in the Middle East, as Thick as Blood," *The Economist*, January 5, 1996.

Watzman, Haim, "Left for Dead," *New Scientist*, February 8, 1997.

The United States

Armstrong, Ellis L., ed., *History of Public Works in the United States, 1776–1976*. American Public Works Association, Chicago, IL, 1979.

Arrington, Leonard J., and Davis Bitton, *The Mormon Experience*, Knopf, New York, 1979.

Barry, John M., *Rising Tide: The Great Mississippi Flood of 1927 and How It Changed America*. Simon & Schuster, New York, 1997.

Bowden, Charles, *Killing the Hidden Waters: The Slow Destruction of Water Resources in the American Southwest*. University of Texas Press, Austin, TX, 1992.

Burke, John, *Buffalo Bill, the Noblest Whiteskin*. Capricorn Books, New York, 1973.

Callahan, North, *TVA, Bridge Over Troubled Waters*. A. S. Barnes and Co. Cranbury, NJ, 1980.

Carothers, Steven W., *The Colorado River Through Grand Canyon*. University of Arizona Press, Tucson, AZ, 1991.

Changnon, Stanley A., ed., *The Great Flood of 1993, Causes, Impacts and Responses*. Westview Press, Boulder, CO, 1996.

Churchill, Beryl Gail, *The Dam Book*. Rustler Printing and Publishing, Cody, WY, 1986.

————, *Dams, Ditches and Water*, Rustler Printing and Publishing, Cody, WY, 1979.

Churchill, Terry S., ed., *The Engineer in America*, University of Chicago Press, Chicago, 1991.

Creese, Walter L., *TVA's Public Planning: The Vision, the Reality*. University of Tennessee Press, Knoxville, TN, 1990.

Davis, Margaret Leslie, *Rivers in the Desert: William Mulholland and the Inventing of Los Angeles*, HarperCollins Publishers, New York, 1993.

Department of the Interior, United States Reclamation Service, *Feature History of the Shoshone Dam*, June 1, 1910.

Douglas, Marjory Stoneman, *The Everglades, River of Grass*. Pineapple Press, Sarasota, FL, 1988.

Douglas, Marjory Stoneman, and John Rothchild, *Voice of the River: An Autobiography*. Pineapple Press, Sarasota, FL, 1987.

El-Ashry, Mohamed T., and Diana Gibbons, *Troubled Waters, New Policies for Managing Water in the American West*. World Resources Institute, Washington, D.C., 1986.

Foote, Stella, *Letters from Buffalo Bill*. Upton & Sons, El Segundo, CA, 1990.

Foster, Mark S., *Henry J. Kaiser, Builder in the American West*, University of Texas Press, Austin, TX, 1989.

Fradkin, Philip L., *A River No More: The Colorado River and the West*. Knopf, New York, 1981.

High Country News, *Western Water Made Simple*, Island Press, Washington, D.C., 1987.

Hundley, Norris Jr., *The Great Thirst: Californians and Water, 1770s–1990s*. University of California Press, Berkeley, CA, 1992.

Jackson, Donald C., *Great American Bridges and Dams*. Preservation Press, 1988.

Karhl, William L., *Water and Power: The Conflict over Los Angeles' Water Supply in the Owens Valley*. University of California Press, Berkeley, 1982.

Lee, Lawrence B., *Reclaiming the American West: An Historiography and Guide*. ABC-Clio, Santa Barbara, CA, 1980.

Leuchtenburg, William E., *The Perils of Prosperity, 1914–1932*. The University of Chicago Press, Chicago, 1958.

Lilienthal, David E., *TVA, Democracy on the March*. Harper & Row, New York, 1953.

————, *The Journals of David Lilienthal: Vol. I, The TVA Years*. Harper & Row, New York, 1964.

Martin, Russell, *A Story That Stands Like a Dam*. Henry Holt, New York, 1991.

Norris, George W., *Fighting Liberal: The Autobiography of George W. Norris*. Macmillan Company, New York, 1945.

Nye, David E., *American Technological Sublime*. MIT Press, Cambridge, MA, 1994.

————, *Electrifying America: Social Meanings of a New Technology*. MIT Press, Cambridge, MA, 1992.

Owen, Marguerite, *The Tennessee Valley Authority*. Praeger Publishers, New York, 1973.

Pisani, Donald J., *From the Family Farm to Agribusiness*. University of California Press, Berkeley, 1984.

————, *To Reclaim a Divided West: Water, Law, and Public Policy, 1848–1902*. University of New Mexico Press, Albuquerque, 1992.

Reisner, Marc, *Cadillac Desert: The American West and Its Disappearing Water*. Penguin Books, New York, 1986.

Reisner, Marc, and Sarah Bates, *Overtapped Oasis: Reform or Revolution for Western Water*. Island Press, Washington, D.C., 1990.

Reynolds, Terry S., *The Engineer in America*. University of Chicago Press, Chicago, 1991.

Robinson, Michael C., *Water for the West: The Bureau of Reclamation 1902–1977*. Public Works Historical Society, Chicago, 1979.

Roosevelt, F. D., *The Public Papers and Addresses of Franklin D. Roosevelt. Vol. 4*. Random House, New York, 1935.

Schlesinger, Arthur M. Jr., *The Age of Roosevelt: The Politics of Upheaval*. Houghton Mifflin, Boston, 1960.

————, *The Age of Roosevelt: The Coming of the New Deal*. Houghton Mifflin, Boston, 1958.

Stegner, Wallace, *Beyond the Hundredth Meridian: The Exploration of the Grand Canyon and the Second Opening of the West*. Houghton Mifflin, Boston, 1954.

Steven, Joseph E., *Hoover Dam*. University of Oklahoma Press, Norman, 1988.

Walton, John, *Western Times and Water Wars: State, Culture, and Rebellion in California*. University of California Press, Berkeley, 1992.

Watkins, T. H., *Gold and Silver in the West*, American West Publishing Company, Palo Alto, CA 1979.

Watkins, T. H., *Righteous Pilgrim: The Life and Times of Harold Ickes, 1874–1952*. Henry Holt, New York, 1990.

Watkins, T. H., and Patricia Byrnes, eds., *The World of Wilderness: Essays on the Power and Purpose of Wild Country*. Wilderness Society, Roberts Rinehart Publishers, Niwot, CO, 1995.

White, Richard, *It's Your Misfortune and None of My Own: A New History of the American West*. University of Oklahoma Press, Norman, 1991.

White, Richard, *The Organic Machine: The Remaking of the Columbia River*. Hill & Wang, New York, 1995.

Whitman, Willson, *God's Valley, People and Power Along the Tennessee River*. Viking Press, New York, 1939.

Wilson, Neill C., and Frank J. Taylor, *The Earth Changers*. Doubleday, New York, 1957.

Worster, Donald, *Rivers of Empire: Water, Aridity & The Growth of The American West*. Pantheon, New York, 1985.

————, *Under Western Skies: Nature and History in the American West*. Oxford University Press, New York, 1992.

Articles

Branscome, James, "The TVA: It Ain't What It Used to Be," *American Heritage Magazine,* February 1977.

"The Earth Movers I," *Fortune Magazine,* August 1943.

"The Earth Movers II," *Fortune Magazine,* September 1943.

Jackson, Donald C., "Engineering in the Progressive Era: A New Look at Frederick Haynes Newell and the U.S. Reclamation Service," *Technology and Culture,* July 1993.

Kusler, Jon, and Larry Larson, "Beyond the Ark: A New Approach to U.S. Floodplain Management," *Environment,* June 1993.

Stegner, Wallace, "Myths of the Western Dam," *Saturday Review*, October 23, 1965.

Steinberg, Theodore, "That World's Fair Feeling: Control of Water in 20th Century America," *Technology and Culture,* April 1993.

Thomas, Dana, "What Henry J. Built," *Barrons,* January 16, 1958.

Zimmerman, Rae, "After the Deluge," *The Sciences,* July/August 1994.

Acknowledgments

Two uncommon young men have been on my mind and in my heart while I was writing: Dr. Goverdhan Singh Rathore, whose heroic resolve on behalf of the people in the villages surrounding Ranthambhore Tiger Preserve is overwhelming, and Wynton Marsalis, because just thinking about all the things he does makes me feel better about being alive—and reminds me to work harder.

I would like to thank the following people:

In the United States: Beryl and Winston Churchill, Lovat Wilkins II and her mother, Jean Cole Anderson, Kay Chernush, Clarice Fabela, David Freeman, Charlie Holland, Dean House, John and Jan Isaac, John Kolars, Sondra Loring, Bill McCormick, Joe Podgor, S. S. Kirmani, Elliot Rebhun, Joyce Starr, Jim Webb, Gilbert F. White, Cal Calhoun, Roland Clement, Cindy Low, Robert O. Collins, Michael Horowitz, Rahul Jacob, Arun Elhance, and the remarkable Ashvin Shah.

In Italy: Maurice Gentilomo.

In England: Tony Allan.

In Australia: Doug Price, Barry Dunn, Julie Bruyn, Murray Jackson, Richard and Amanda McCarthy, Alan Frost, Jack Grimstead, Kathy Manthex, Alan McHardie, Tania Smith, and J. D. Blackmore.

In Spain and Portugal: Pedro da Cunha Serra, José Manuel de Palma, and Manuel da Silva Paraira.

In Pakistan: Attia and Ayub Qutub, F. A. Zuberi, Irshad Ahmad, Bapsi Sidhwa, Ahmjad Agha.

In Palestine and Egypt: Sharif S. Elmusa.

In Yemen: Ruth Roberts, Musaffer Erselçuk.

In India: Fateh Singh Rathore, Avani Patel, Usha Singh Rathore, Balram
    Singh, Raghu and Meeta Rai, Diwann Manna, Patwant Singh, B. B.
    Vohra, B. G. Verghese, Shadiram Sharma, and V. K. Nanda.

In Holland: Co Van Dixhoorn, Pieter Huisman, Louis Van Gasteren, G.
    Te Slaa, Anja Nijenkamp, Willem Zwigtman, Rajiv and Jotsna
    Mehra, Louis and Sophie Paul.

In Turkey: Olus Arik, Ali and Sevin Balaban, Ayse and Ahmet Oncu,
    Sinan Kelali, Nimet Osguç, Bulent Gultekian, Kamran Inan,
    Ismail Polot.

I am exceptionally favored to have Gerry McCauley as my literary
agent. This book wouldn't exist without him. Personal thanks to Chris
Knutsen, Julie Grau, Tom H. Watkins, Dr. Jody Davies, Rasil Basu,
Luye Lui, Joan & David Grubin, Cal Raines, Martha Saxton, Enrico Fer-
orelli, Andrew Ward, and the exquisite Susan Vale Gibson.

My behen, Sajni Thurkral, in New Delhi and Mira Balram Singh at Tiger
Haven Farm fussed over me worriedly, as did my pal Bob Strozier just
down the street. Special thanks to my beloved father- and mother-in-law
Champ and Dewey Ward, and to my lovely stepson, Nathan Ward, and
my dear daughter-in-law, Katie Calhoun. My beautiful daughter, Kelly,
did translations from the French. Thanks to my mother, Helen Raines,
and my father, who lovingly read each chapter, especially my dad, who
didn't make it to the end.

The incomparable Sheila Lawton traveled with me from Australia to
Audalusia, and the magical writer and artist Miles Gibson coddled me
endlessly, from an ocean away. My wise, supportive son, Garrett, contributed paperwork, legwork, and thoughts. But above all I want to
thank my seriously loving husband, Geoff, without whom I could do
nothing, nothing at all.

Index